Andreza Araújo

# SAFETY CULTURE

From Theory to Practice


2st Edition

São Paulo - 2022

General Coordination

Andreza Araújo

Graphic Design and Cover Designer

Reginaldo Saw

Translation

H3 Traduções



Araújo, Andreza Moleiro
Safety culture : from theory to practice / Andreza
Moleiro Araújo. -- 2. ed.
280 p.

Santana de Parnaíba, SP : Ed. da Autora, 2022.

Bibliografia.

ISBN 978-65-00-44718-7

1. Cultura de segurança 2. Engenharia de segurança

I. Título.

22-110169                               CDD-620.8

Índices para catálogo sistemático:

1. Cultura de segurança : Engenharia 620.8

Eliete Marques da Silva - Bibliotecária - CRB-8/9380

# Summary



# Appendix

To my little son Tito, an extraordinary gift that I received, may we continue to grow together; and that together we can continue to grow.

# Acknowledgment:

A writer I admire once said that writing a book and crossing a desert are very similar things. After all, large expanses of dry, flat land are occasionally dotted with wells of inspiration. This is a small thank you to dear friends who made this journey so enjoyable. Thank you for not complaining while we were on the sand and for celebrating when we were at the oasis.

My journalist partner, Ana Paula Alfano. You do more than organize and edit what I write. You preserved my genuine message.

My team, for their contributions and ideas.

To my dear husband. Nobody can do what you do the way you do it. We are indebted to you.

Helena and Tito. What mother deserves such wonderful children? (count on me whenever you want).

To my parents for their lifelong friendship.

And a prayer to the giver of all words, thank you to the Eternal, my Master and personal friend, Jesus Christ, for the journey and the call to always make a difference, practicing care with a strong and genuine message.

And to you, the reader, my final salute. For some of you, this book marks our fifth meeting (happy birthday!). For others, this is our first (nice to meet you!). For most, we're somewhere between the two ends (good to see you again!).

You're about to give me one of your most precious possessions: your time. I'll do anything to be a good butler. While writing a book can be similar to a journey through the desert, the act of reading a book should not be. It must be a rest stop in an oasis. I hope that's the case with the following pages.

# Introduction

*Care is contagious.*

Andreza Araújo

When it comes to accidents at work in Brazil, companies often still focus excessively on numbers. Indeed, they are important. In Brazil, every 48 seconds there is an accident at work. This means that, in a single 8-hour shift, 600 workers do not return home and to their families at the expected time. To make the statistics even harsher, every 3 hours and 38 minutes there is a fatal work accident. With these data, Brazil currently occupies the 4th place in the World Accident Ranking (1). And the data from Latin America are even more alarming - an employee suffers from an incident every 15 seconds.

When I think about the huge number of people impacted, their families and the lack of a safety culture within companies, I am very concerned. Brazil has an army of disability retirees, about 3,470,0001, excluding death cases.

It should be noted that these data refer to workers who have a work registration.

But while the numbers say something about safety, it's critical that we look beyond them. Numbers are merely indicative; they are not able to explain the reasons behind each accident. One needs to understand the root cause that leads to the occurrence. Why, in the face of dangers and risks within the work environment, do workers not choose or see safety as the most important alternative? What makes them deviate from correct and safe action?

I usually say that an accident at work is a book we don't read. We know that it's not a surprise event, it never occurs without warning. Before materializing with more serious damage or injuries, there are several warnings – and this is proven by the various models of Deviation (or safety) Pyramids known and used for decades in the area of accident prevention. The most famous of these is Bird's Pyramid, developed in the 1970s by the North American engineer Frank Bird and which we will discuss later in this book, the result of an analysis of about 1.7 million accidents reported in 297 different companies. According to Bird, for every serious or fatal physical injury there are 10 minor physical injuries, 30 material damages and 600 near misses. It is, therefore, an absurd amount of alerts that ends up being ignored on a daily basis, both by workers and by their leaders, certainly due to the absence of a safety culture within the company. Occurrences are therefore multifaceted events.

Brazilians are a people known for improvising solutions. Laws and rules often exist, but with a certain permissiveness, they break them. Listen, I don't intend to impose any judgment here, mainly because, anthropologically, one of the roots of this improvisation habit of ours is emotion – yes, our heart! Brazilians often act driven by emotion rather than reason and have developed a kind of historical propensity for informality. In everyday life, and also within the work environment, it is common, for example, to break or merely ignore the rules in favor of friendships. Being willing to help friends is a noble thing, but not when it comes with breaking the rules and considerably increasing the risk of accidents.

Another reason is that, according to studies conducted over the past few decades that have deepened our understanding of how our brains work and how people make decisions, we humans are not perfectly rational decision makers. Our thinking system is divided into two parts, one logical and the other intuitive. Logic takes more energy and time. Intuition comes faster. In the course of a single day, according to these studies, we make far more intuitive decisions than logical ones. We therefore follow an apparently irrational behavior in risk management. Israeli psychologists Daniel Kahneman and Amos Tversky, two of the most influential economic thinkers in modern history, are the fathers of these behavioral theories, which indicate how error-prone human decision-making is.

But what sounds like a bad thing can actually be an advantage. Because, if we know in advance that there are chances of making mistakes in moments of decision, we are also able to anticipate and even quantify the flaws in our thought processes [*]. This theory by Kahneman and Tversky has already been widely applied, with success, to economics, the legal environment, marketing, and public policy, but we are only now beginning to understand how it impacts and can also be used in safety.

If safety is a result of decision-making, consciously identifying and considering risk factors before needing to act (intuitively) greatly reduces the chances of something going wrong. That's why it's so important to transform safety rules into a safety culture. Procedures must reflect the correct way to perform each activity, and safe behavior must become a habit. Decisions cannot be made "in the heat of the moment," because the chances of them not being the right ones are huge. Hence it's easy to understand that the frightening accident numbers we have today reflect our unstructured safety learning. We don't learn about safety at school. Where are we learning about safety? We are learning within companies that are chasing safety as their first value.

Many companies and methodologies already see cultural transformation as the main way to strengthen safe habits. We know that culture is strengthened through symbols, rituals, heroes, and communication. We have many tools to explore and stimulate. I don't like to prioritize them, I prefer to think of a model that is a kind of solar system of safety, in which each tool contributes in a unique way to achieving a sustainable and interdependent culture.

This book undertakes to explore this solar system, with each planet being an important tool. We will, from the traditional H&M and DuPont (Bradley Curve) models, propose a hybrid and functional model. For the construction of this model, we started the journey with Frank Bird and his pyramid, which brought us the contribution that no accident is born catastrophic. In fact, each accident is a book composed of several chapters and the last one in Frank's conception is the near miss. Years after Bird's publication, a new statistic was added to the base of his pyramid - that for every serious or fatal accident, there are about 30,000 misbehaviors or unsafe acts committed. This number, the result of an extensive study by the DuPont consultancy carried out in the 1990s, showed that indirect safety expenses within organizations ended up being five times greater than direct costs, an information that cannot, and should not, be ignored.

In the following pages, based on my experience in studies and projects carried out in more than 30 companies in the last 17 years, training more than 15 thousand leaders in 25 countries, I will present other levels that I consider fundamental when evaluating the causes of a serious or fatal accident, and their relationship with building a solid and perennial safety culture. They are: behavioral triggers, beliefs, values, culture, and the immediate leader's role.

Starting with the behavioral triggers that explain the deviations in behavior and concern the thoughts, feelings, and emotions of our employees. They are conventionally categorized into four different groups – social, cognitive, psychological, and physiological. Every step we take is influenced by one or more of these factors, as we'll see throughout this book. These are the triggers that need to be addressed and dealt with in order to achieve the materialization of care, breaking old and ingrained paradigms about safety that workers – and almost always their leaders as well – still carry in their work environments.

One needs to develop a keen awareness that safety, more than anything, is about people. And if we humans have care as a basic instinct, when this instinct is genuinely triggered and stimulated we are able to create a care chain throughout the organization.

It is necessary to understand that the impact that behavioral triggers will have depends on the deeply rooted truths that we call beliefs. What we believe defines how we will respond to behavioral triggers. For example: when I come to work leaving my mother in the ICU, can I be aware that worries and emotions can cause me distractions? Do we have limiting beliefs or empowering beliefs about safety to face a situation like this?

Beliefs reflect our values, that is, what we do not negotiate and do not accept to live without and/or lose. In April 2019 I ran an experiment on all the leader trainings I executed that month. At one point in the Workshop I asked participants to take a piece of paper and write the names of six people, pets, attributes, e.g., character, dignity, health, faith, physical integrity etc. Names of people and things they couldn't live without, that is, the non-negotiables of their lives. After they all finished writing I asked: Who wrote their own name on this list?

The absolute answer was nobody. Do you see the inversion of values? We say that safety is care, that it must start with us, however, in practice this concern and attitude is not reflected in the values, taking care of yourself is still on the scale of important things.

Can you see the importance of that? Beliefs and values make up our culture and will be reflected in behaviors and habits. Which can be perceived in our choices and attitudes.

With a simple and innovative approach, I intend to present and discuss these themes and how safety humanization has brought the individual as a protagonist, and can be worked in a transversal way, going beyond traditional safety models.

The dream is that the safety we preach and practice within corporations will have enough energy to reach our employees' dinner table, to impact their families. It must go far beyond rules nailed to the wall. Safety needs to be re-signified as Care and Care needs to be practiced on 3 different levels: First, I need to take care of myself; then I extend care to my neighbor; And lastly, I also allow myself to receive someone's care.

May we be the catalysts for a strong and perennial safety culture, transforming the lives of our employees with the habit called "Care"!

## Brazil: scenario and context

Talking about accidents at work is talking about people. Therefore, it is essential, in the search for an understanding of the root causes of these episodes, to look deeply into the scenario in which our employees are inserted. There are many problems that currently afflict the country, the economy, families, and that can interfere with income and attention in the work environment. Imagine the concentration of a parent who leaves a drug-addicted child at home. Or the pressure of a worker who has an unemployed partner and whose livelihood depends exclusively on their salary. There are many factors that can influence and lead to an accident. Below are some numbers referring to some of them.

## Depression

The main cause of health problems and disability worldwide, it affects more than 300 million people. According to data from the WHO (World Health Organization, 2015), we are the country in Latin America with the highest prevalence of the disorder - it affects about 11 million Brazilians - a number equivalent to the total population of Greece or Cuba. In addition, the country holds a world record in the prevalence of anxiety disorder (9.3% of the population, about 18 million people, or one in 11).

Sadness, lack of energy, loss of interest, reduced sleep, loss of concentration, restlessness, indecision and even suicide risks – all these possible symptoms of depression affect the worker's routine. In addition, it is necessary to consider that, with these high rates of depression in the country, there are chances that, if it is not them facing the disorder, it will be someone close to them, which can also interfere with work.

## Occupational depression

More specifically, a depression caused by the work itself also tends to greatly affect decision-making. Being constantly frustrated, stressed, not getting enough rest or adequate working conditions, and feeling like you're not valued or recognized can trigger depressive symptoms. The worker's mental health is a determining factor for the worker's safety and the rest of the team's. Therefore, to prevent work-related mental and behavioral disorders, it is necessary to develop a wellness agenda.

## Serious illnesses and deaths

Imagine having to go to work and operate a machine after receiving a diagnosis of a serious illness - yours or a close relative's - or after the death of a loved one. You cannot expect the same concentration and decision-making power from a worker. In 2015, Brazil recorded about 210,000 deaths from cancer and 350,000 related to cardiovascular diseases. If we compare that with data from 1998 – almost 20 years ago –, it is a 90% increase in mortality from neoplasms and 36% in mortality from cardiovascular diseases.

## Chemical dependency

According to data from a survey by the Ministry of Health (from 2016), 19.1% of the Brazilian population has abusive consumption of alcoholic beverages (27.3% in men and 12.1% in women). In the last ten years (from 2007 to 2017), alcohol consumption in the country increased by 43.7%. Alcohol is the most consumed licit drug in Brazil, but it is also necessary to consider the problems that many Brazilian families face with illicit drugs. IBGE data show that the use of illicit drugs in the country – cocaine, crack, opioids – jumped from 0.8% to 7.3% in the last decade. Among female adolescents aged 13 to 17, the users' rate went from 6.9% to 9.2% in this period. Can you imagine having to go to work after having hospitalized a daughter or son for chemical dependency?

## Unemployment

Data from IBGE's continuous National Household Sample Survey (Pnad) point to an unemployment rate in Brazil of 13.1% in the first quarter of 2018. This represents about 13.7 million unemployed, slightly less than the total population of Bahia, the fourth most populous state in the country.

## Family violence

Try to put yourself in the shoes of someone who left home for work after being beaten up by their partner. Or that had to face a shift knowing that their own daughter goes through this kind of situation. Men are also subject to this category of crime, but women are still the vast majority of victims. In Brazil, in 2017, around 193,000 women filed a complaint for domestic violence, according to data from the Brazilian Public Safety Forum. An average of 530 women activate the Maria da Penha law per day, that is, 22 per hour.

Not to mention the number of rapes recorded. Brazil has about 135 new cases of rape per day, which represents 50,000 reports per year, according to the Violence Atlas (2018). Considering that the vast majority of victims do not report this type of crime – the rate of underreporting, almost always out of fear or shame, is so high that only between 7.5% and 10% of them are reported to the police – the total may reach 500,000 rapes a year.

Data from the Digital Observatory of Occupational Health and Safety (from the Public Ministry of Labor), referring to the period from 2012 to 2017, in Brazil:

Social Safety Expenses with Accidental Benefits:

# R$ 75,523,000

**4,403,906** estimated accidents from 2012 to today.

**1** estimated accident every 48sec.

**16,373** estimated accident deaths.

**1** accident death estimated every 3h 38m 43s.

# What We Know About Safety Culture

*A man awakened by the meaning of safety will awaken a second man, who will go in search of a third. Three awakened men can transform the safety value of an entire factory.*

Andreza Araújo

The night of April 26, 1986 was supposed to be a routine test at Reactor 4 on one of the buildings at the Chernobyl nuclear power plant, 130 kilometers north of Kiev, Ukraine, the then Soviet Union. The engineers wanted to know how long it would continue to supply power after several power outages. Such a test had already been carried out a year earlier. But this time, to measure time accurately, they would need to disable the reactor's automatic shutdown mechanism. What followed was, to this day, over 30 years later, the largest nuclear accident in world history. An explosion and a series of fires released fuel and radioactive material into the atmosphere. Two workers died at the time of the first explosion, another 28, including firefighters, in the following weeks from poisoning. In the first few months, hundreds of people were diagnosed with radioactive iodine contamination. The Soviet government had to move 120,000 people away from the worst-affected area in the first few hours, and another 240,000 in the following years. Still, there were hundreds of thyroid cancer records.

Despite the catastrophe of unimaginable proportions, a report issued in 1991 that analyzed all the actions of the engineers involved in the episode concluded that, even though they made the reactor unstable and took the explosion for granted, they had not violated any operating policy or safety principles. Simply because there were no policies or principles in place. And, even if they were nailed to the wall, it is known that they would not have prevented the event. That's because the causes of the explosions at Chernobyl were not just technical. They were also, and mainly, cultural. And, until that fateful night, almost nothing was discussed, around the world, about safety culture.

The Chernobyl plant was built by the Soviet government to produce an atomic bomb, regardless of the possible costs of doing so, including human lives. Most of the workers came from the Gulag, forced labor camps to which criminals, political prisoners and any citizen who was against the communist regime were subjected. Those who participated in the construction of the most secret sectors of the plant were later isolated for the rest of their lives, like the physicist and Nobel Peace Prize laureate Andrei Sakharov later described in his biography. It is possible to imagine what it was like to work in a place like that, knowing what the future held. Even if such safety rules existed and were widespread, they would be the last thing a worker would worry about.

The Chernobyl accident and its worldwide consequences led, in the following years, the International Atomic Energy Agency (IAEA) to create the concept of a SAFETY CULTURE – a culture that influences the organization's structure and style, as well as its attitudes, approaches and the individual commitment of employees at different levels.

But before we talk about safety specifically, let's think about the broader concept of culture. In the Houaiss dictionary, culture is defined as a "set of patterns of behavior, beliefs, knowledge, customs, etc. that distinguish a social group; form or evolutionary stage from intellectual, moral, spiritual traditions and values."

It is to the group what memory is to individuals, including traditions that worked in the past and assumptions about how the world and people are, think, and should act.

In organizations, culture is also a set of behavior patterns, generally implicit and shared by everyone, formed from their values and beliefs. It is the result of the interaction between the people of the company, within that same environment – and the environment itself ends up being, also, a result of this interaction. A widespread example is that a corporation's strategies represent the bricks, but the mortar that keeps the walls solid and upright is the company's culture. Therefore, the more fragile it is, the greater the risk of the wall collapsing.

To me, a strong safety culture is one in which each individual sees safety as a non-negotiable value, and not just as a priority, obligation, or goal to be accomplished. Therefore, everyone is responsible for it, at each hierarchical level within the corporation. The key lies in making the practice of care an individual and institutional issue, both at the same time.

More specifically, the book Elements of the SMSQRS Management System - Vulnerability Theory lists safety culture concepts from different authors and institutions:

**International Atomic Energy Agency (IAEA, 1991):** "Set of practices and attitudes established within organizations and individuals, prioritizing attention and importance to the safety aspects of nuclear installation";

**Uttal (1983):** "Shared values and beliefs that interact with the organization's structure and control systems affecting the pattern of behavior";

**Turner, Pidgeon, Blockley & Toft (1989):** "establishment of beliefs, standards, attitudes, rules and socio-technical practices concerned with minimizing the exposure of workers, managers, customers and the community to potential risk situations that may result in injury";

**Britain's Nuclear Installation Safety Committee (1993):** "results of values, attitudes, training and behavioral requirements, individual and collective, that make it possible to seal the commitment with the form and the search for effectiveness of the HSE programs of the organizations."

**Confederation of British Industries (CBI, 1991):** "ideas and beliefs that all members of the organization must share about risks, accidents and harm to health";

**Carnino (1989), Lee (1993) and Lucas (1990):** "Organizations focused on a proactive safety culture are characterized by a communication system based on mutual trust, exchanges of internal perceptions on the importance of safety values and confidence in the effectiveness of the preventive measures adopted."

**Energy Institute:** safety culture is formed by a group's core beliefs and values regarding risk and safety. It is, informally, the "how things are done around here" and, for them, when it is strong, everyone:

- sees safety as a value;

- is alert to expect even the unexpected;

- fully understands what they must do in the name of health and safety;

- is open to new ideas that improve health and safety; and

- wishes to make a difference and believe that their behavior makes a difference to others

As you can see, the concepts have slight variations. And they are unanimous in citing values, beliefs, and behavior as the safety culture axis. For me, safety culture is a set of values and beliefs translated into behaviors that become habits and are perceived through rituals, symbols, heroes, and communication. In practice, it is not an independent action that starts from the assumptions of the CNPJ (Corporate Taxpayer Registration), it starts from the CPF (Individual Taxpayer Registration), it is personal, individual and when it infects the collective it can move the CNPJ (Corporate Taxpayer Registration).

And where do organizational culture and safety culture meet? Aspects of organizational culture have an impact on the attitudes, behaviors, beliefs, and perceptions of the entire team, including those related to increased or decreased risk of accidents, general health, and safety (Guldenmund, 2000). All organizations develop their own cultures because leaders share similar values and attitudes. And these values and attitudes also guide how members deal with safety and prevention, thus reinforcing them positively or negatively within the organization. In short, the way safety is perceived, valued, and prioritized reflects the strength of a company's safety culture, and becomes part of its organizational culture.

## Safety Culture is an integral part of all types of culture:

**Informational culture:** values reporting, based on the elements of the safety management system.

**Flexible culture:** adaptable to facing risks and dangers.

**Learning and anticipation culture:** encourages the will to learn the right lessons and implement solutions, seeking where the risks are and understanding how they manifest themselves in order to carry out the systemic addressing of preventive actions.

**Engagement and collaboration culture:** encourages workforce participation in which people support and care for one another.

**Fair culture:** makes it very clear what acceptable and unacceptable behaviors are and has consequence management to deal with non-compliance with the rules.

## Safety culture operation

Culture is always a reflection of people's beliefs and values, something that becomes visible from their choices and behavior. Let's compare the way it works to an onion, with its many layers, from the innermost to the outermost.

In the heart of the onion are our beliefs and values. From our beliefs and values arise three layers from which it is possible to see the presence and operation of the Culture. The first layer is composed of the rituals, the second by the symbols, the third by the heroes of the organization. Transversal to these layers are the communication practices that permeate all the others. The figure below idealizes the model that will be further explained.

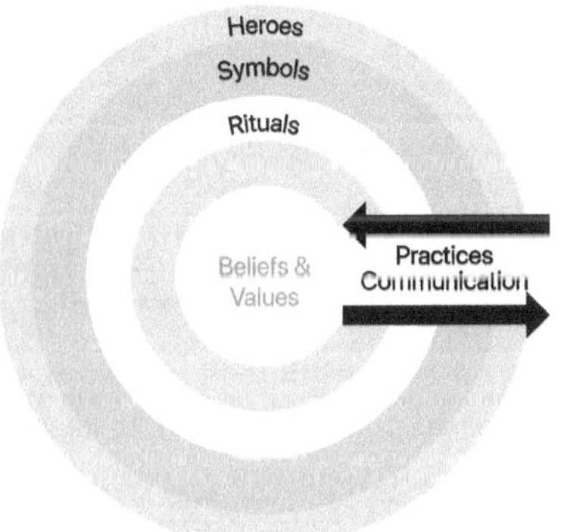


Figure 1: Onion model to understand the culture.

The first layer, which is also the innermost, is formed by rituals, those events with their own rite and defined periodicity that reinforce safe behaviors and practices. As an example, we can mention DDS rituals, committee meetings, the CIPA process, inspections, checklist, and Work Permits. To make this example even clearer, let's think about this same layered model applied to the Christian religion. Figure 2 compares the Christian religion and safety. When we make the parallel with Christian culture, we have faith at the center and as the main ritual we can mention religious cults.

The second layer is the symbols that reinforce the commitment and presence of safety, through visible artifacts, including PPE, signaling and isolation boards, boards with KPIs and other communications. Again, when we make the comparison we can have the Bible, crucifix, and images of saints.

The third layer refers to the heroes, who can also be called leaders, highest authority, executive committee, those who show where to go safely. These are the examples of these heroes who end up becoming the rules of conduct, each organization has its own. In religion we have Jesus as the central hero of Christian culture.

Finally, there is a dynamic transversal layer that talks to all the others, aligning them in order to create an environment of trust, coherence and constancy of safety, this layer is called communication practices.

When we understand how culture works, it is possible to create stimuli and direct tools so that, associated with rituals, symbols, heroes, and communication, they can increasingly strengthen the Safety Culture, so that it becomes a value and influences the lives of our employees in an integral way, that is, anytime, anywhere and with anyone.

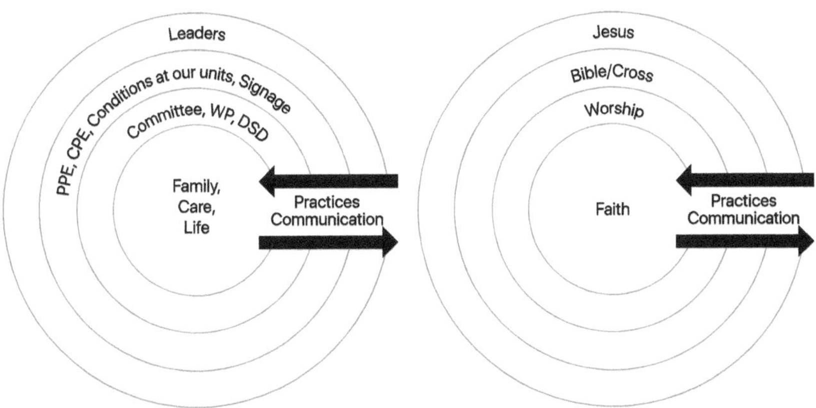


Figure 2:  Examples of how the onion model works.

# Indicators of a strong safety culture

After an accident between two trains at Landbroke Grove Junction in London, England, which, in October 1999, left 258 people injured and caused 31 deaths, in addition to all the material loss, an investigation was carried out to define the causes and what should be changed so that it wouldn't happen again. The final report listed the top five indicators of a strong and efficient safety culture:

a)   **Leadership:** it must be clear, decisive, and present in the entire team's routine. It must serve as an example and, therefore, experience what it preaches and demands from its employees. We'll talk more about leadership in a later chapter in this book.

b)   **Communication:** one of the pillars of efficient management is communication. The leadership must send clear messages to the entire team regarding the expected goals and attitudes towards safety; at the same time, employees must provide constant feedback to their leaders. Communication, therefore, must be bidirectional.

c) **Employee involvement:** it is not possible to achieve a satisfactory degree of safety if employees do not feel responsible if they are not constantly motivated regarding their importance in their own health and that of their colleagues.

d) **Continuous learning:** implementing improvements and new rules and not following up on them and developing them is of little use. Organizations and their employees need to learn from previous accidents or near misses (including those that happen in other companies), analyze information and data, understand changes in behavior that arise over time.

e) **Attitude towards guilt:** it is necessary to have zero tolerance for what is not safe. Any and all information regarding this, even that which makes a colleague responsible for something, must be reported without fear of retaliation or recrimination. Guilt is often justifiable and dealing with it fearlessly is a step up to a safer company.

## The difference between safety climate and culture

When talking about safety culture, both organizational and individual factors must be considered. What is the perception of each employee – across the hierarchy – about the risks and what is their involvement in decision-making related to safety? What is their influence on the rest of the team? And how does the leadership and team responsible for safety architecture within the corporation relate to this information?

Over the past two decades, several experts have delved into research and attempted to develop and test techniques capable of measuring the climate and promoting a positive safety culture within corporations. It is important to differentiate between climate and culture.

Climate is just a temporal manifestation of organizational culture, being, therefore, more flexible, and changeable. And easier to measure. Culture itself is characterized by more ingrained aspects that need time to be transformed.

And how, then, to measure safety climate and culture, taking into account that they are a sum of organizational and individual factors? A framework published by T.D. Jick (1979) structured a multiple perspective model built on three different foundations. A model that, to this day, decades later, is used by different authors and in many organizations. It is:

1) Attributes of the organization (what it is or already has) or situational aspects.

   △ Manifested in: safety policies, systems and processes, frameworks, reports.

   △ Methods used to measure: observation, audits, etc.

2) Perceptions of the organization (how it is seen). It is the "safety climate," linked to psychological aspects.

   △ Manifested in: employees, suppliers, external public in general.

   △ Methods used to measure: interviews, surveys, etc.

3) Individual perceptions (impact on people) or behavioral aspects.

   △ Manifested in: commitment, attitudes, behaviors, and responsibilities of each employee.

   △ Methods used to measure: questionnaires, observation, etc.

## Metrics available to measure safety culture maturity

Culture is, by definition, something in constant motion and somewhat abstract. Therefore, it is difficult to quantify or qualify it in a standardized way. To facilitate this process, globally recognized metrics are used as models – one from Hearts & Minds (Energy Institute) and another from DuPont known as the Bradley Curve. The two are similar, with minor differences in their representations.

The Hearts and Minds model was developed by Shell E&P, based on 20 years of university research, and is being successfully applied in Shell and non-Shell companies worldwide. Hearts and Minds uses a variety of tools and techniques to help an organization engage all employees in managing workplace safety as an integral part of its business.

## H&M Curve 1986

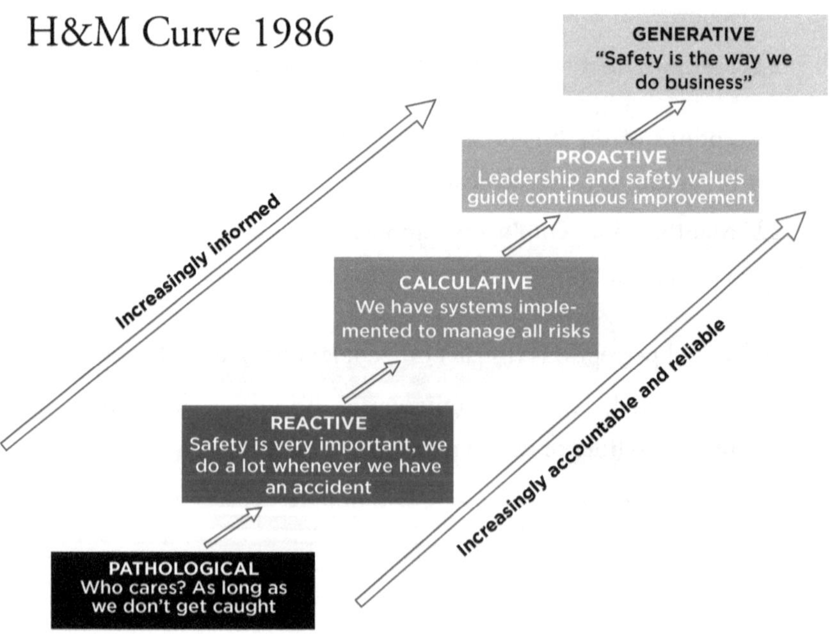


Figure 3:   Hearts and Minds Model

## Pathological



If they could, people wouldn't even think about safety within the organization. They only act according to the rules because they have to. Therefore, there are no effective actions in the area of occupational safety within the corporation. Legislation is complied with, period. It is a stage of safety denial.

## Reactive



The topic is taken seriously, but only because something has gone wrong, some accident or incident has left its mark. Actions are not systematic, they seek to respond only to work-related accidents, focusing on remedying and not preventing them. Safety leaders feel frustrated when they see that the team does not act in accordance with what they have been taught and/or requested, regarding safety.

## Calculative or Bureaucratic



Here too much importance is given to systems and data. Numbers are collected and analyzed exhaustively, there are audits and people start to feel that they already understand how things work. However, the efficiency of such systems and collected data often leaves something to be desired. So far, culture is still seen as a priority and competes with other business items. Another typical phenomenon in bureaucratic cultures is that safety is an entity for which people do not feel responsible. It's just one topic or department.

## Proactive

*"Here, leadership and safety values are constantly evolving."*

The organization does not live with its eyes on past incidents, but on the prevention of future ones. Leaders, based on the company's strong safety values, drive continuous improvement, and seek to anticipate problems, considering variability. The team, more than executing safety rules, is focused on giving new ideas and making them more efficient every day. At this level, safety is recognized as a value. Safe behaviors and habits go beyond factory walls.

## Generative or Sustainable

*"Health and Safety is how we do business here!"*

The organization already has a high level of commitment but aims to constantly improve it. Failures serve as a stimulus for this, not to point out culprits. Information, communication, and engagement are keywords. There is an integrated system on which the company relies to conduct its business and always find the best ways to control and anticipate risks. In addition, these themes are fed back by what happens at home.

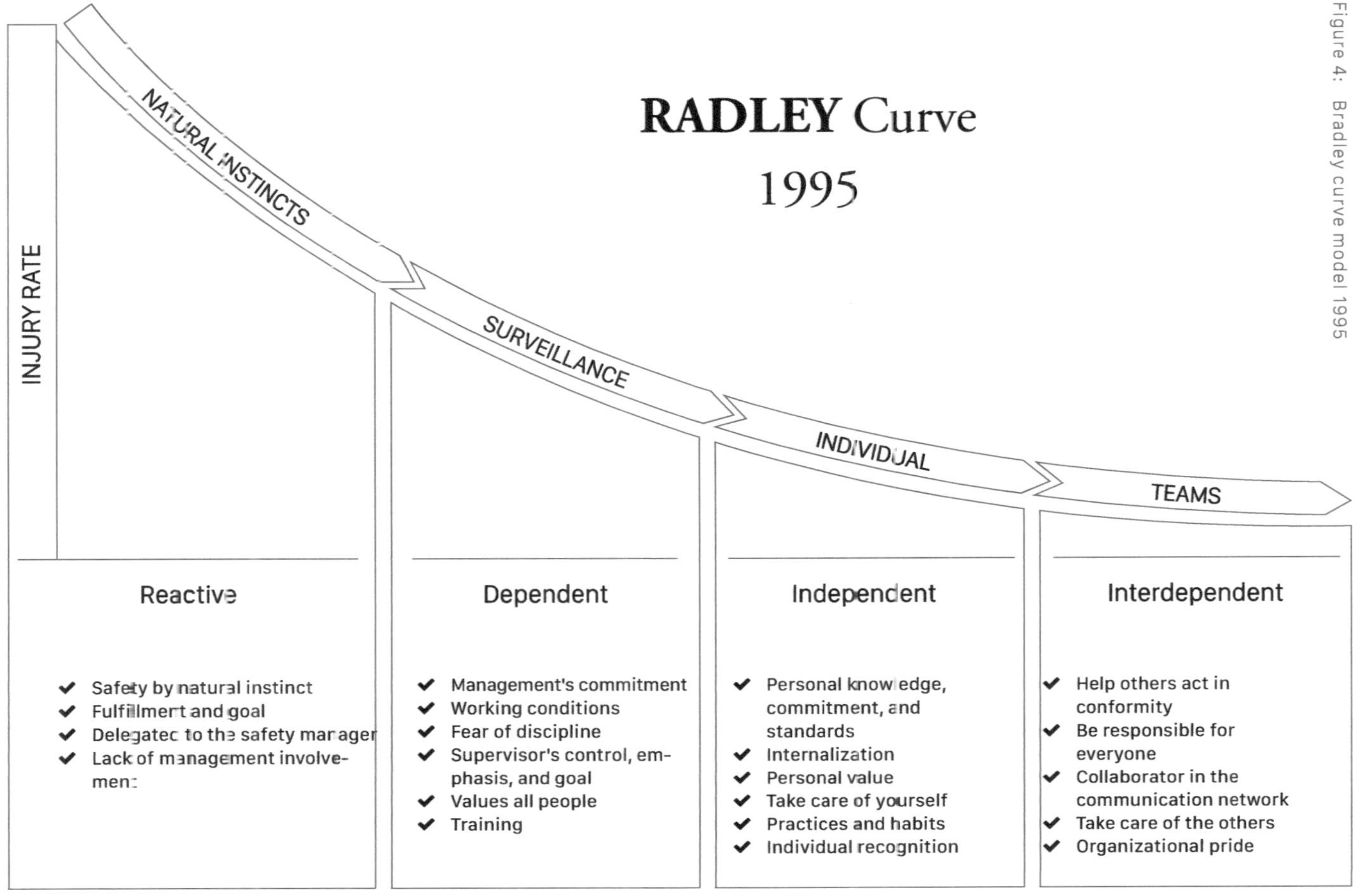
RADLEY Curve
1995
INJURY RATE
NATURAL INSTINCTS
SURVEILLANCE
INDIVIDUAL
TEAMS
Reactive
Safety by natural instinct
Fulfillment and goal
Delegated to the safety manager
Lack of management involve-
ment
Dependent
Management's commitment
Working conditions
Fear of discipline
Supervisor's control, em-
phasis, and goal
Values all people
Training
Independent
Personal knowledge,
commitment, and
standards
Internalization
Personal value
Take care of yourself
Practices and habits
Individual recognition
Interdependent
Help others act in
conformity
Be responsible for
everyone
Collaborator in the
communication network
Take care of the others
Organizational pride

The Bradley Curve, a model patented by DuPont based on studies that began in the 1990s, lists four stages of safety culture maturity: reactive, dependent, independent, and interdependent.

**Reactive:** linked to the natural instincts of each employee, who does not assume responsibility and sees safety more as a matter of luck than risk management and control. For them, "accidents merely happen." And, obviously, with this kind of behavior, they do happen.

**Dependent:** safety is delegated to supervisors; it is seen as a matter of following rules previously developed by third parties. Accident rates tend to decrease and the team responsible for safety and risk management believes that accidents could be avoided "if people followed the rules."

**Independent:** you begin to look at yourself as responsible for your own safety. Employees come to believe that their actions can make a difference. Accidents decrease even more.

**Interdependent:** safety is something that depends on everyone. In a mature safety culture, it becomes sustainable, with accident rates close to zero. Everyone feels capable and authorized to act as necessary to ensure their own and others' safety. Care is practiced at all levels.

The differences between the H&M and DuPont models are in the graphical representations and terminology used. While DuPont lists four maturity levels, H&M proposes five – DuPont doesn't recognize the pathological level, that of safety denial. H&M works with 23 safety culture-enhancing dimensions and DuPont with 22 elements.

Next, I compare the two models to a third one, built by me over the last few years, showing which are the safety culture's strengthening elements. These elements are reflected in rituals, symbols, in the behaviors and attitudes of Heroes and in communication practices.

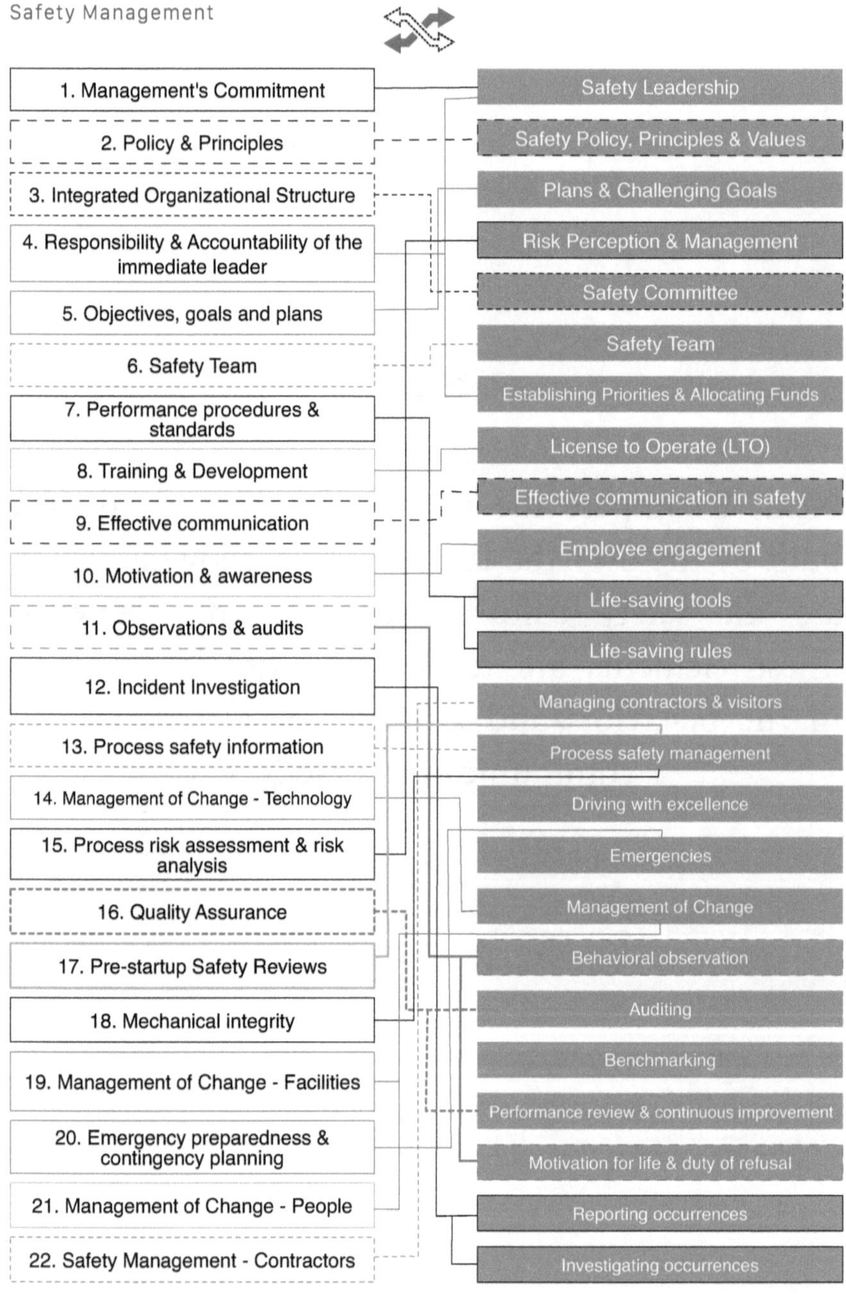

22 Elements
Dupont (Culture + PSM) - Process Safety Management

24 Elements
Andreza Araújo

1. Management's Commitment
2. Policy & Principles
3. Integrated Organizational Structure
4. Responsibility & Accountability of the immediate leader
5. Objectives, goals and plans
6. Safety Team
7. Performance procedures & standards
8. Training & Development
9. Effective communication
10. Motivation & awareness
11. Observations & audits
12. Incident Investigation
13. Process safety information
14. Management of Change - Technology
15. Process risk assessment & risk analysis
16. Quality Assurance
17. Pre-startup Safety Reviews
18. Mechanical integrity
19. Management of Change - Facilities
20. Emergency preparedness & contingency planning
21. Management of Change - People
22. Safety Management - Contractors

Safety Leadership
Safety Policy, Principles & Values
Plans & Challenging Goals
Risk Perception & Management
Safety Committee
Safety Team
Establishing Priorities & Allocating Funds
License to Operate (LTO)
Effective communication in safety
Employee engagement
Life-saving tools
Life-saving rules
Managing contractors & visitors
Process safety management
Driving with excellence
Emergencies
Management of Change
Behavioral observation
Auditing
Benchmarking
Performance review & continuous improvement
Motivation for life & duty of refusal
Reporting occurrences
Investigating occurrences

# 24 Elements
Andreza Araújo

# 23 Dimensions
Hearts & Minds

FROM THEORY TO PRACTICE

# Safety Culture's Strengthening Elements

*"Excellence is an art won by training and habituation. We do not act rightly because we have virtue or excellence, but we rather have those because we have acted rightly. We are what we repeatedly do. Excellence, then, is not an act but a habit."*

Aristotle

**There are 24 fundamental elements** for corporations that seek to strengthen a safety culture. It is necessary to measure the organization's adherence to each one of them in order to be able to classify its culture as pathological, reactive, bureaucratic (or calculative), proactive or sustainable. Throughout this book, we will delve into some of them, but in this chapter I list and explain what each one represents.

As we go through each element, I want you to reflect on how it can manifest itself, through rituals, symbols, in the daily lives of the organization's heroes (spokespersons) and in communication practices.

Also, add to this reflection that making all this visible goes beyond conventional Walk The Talk. It is necessary: to speak, to do and to believe, that is, we need the genuine that includes the values, individual beliefs that, when added together, leverage the organizational climate, and move the culture.

In practice, we can see the 24 elements working in a dynamic PDCA format - Plan, Do, Check and Act, Figure 5, brings this operation proposal. Sorting elements by their Planner, Executor, Check and Action potential.

I consider this view interesting, since each element needs to seek continuous improvement, we cannot admit that safety culture is treated in a static way. Furthermore, if we consider that we started at the CPF (Individual Taxpayer Registration) towards the collective, it is necessary to think, work and develop in a systemic model, encompassing process and management that becomes a personal value. People need to be convinced, believers, filled with purpose that "Care" is our first and most important habit for life and that practicing it is a constant challenge.

To bolster this argument, I bring up the recent study - Beyond Performance 2.0 (John Wiley & Sons, July 2019), company executives who put time and effort into dealing with individual mindsets were four times more likely to "succeed" than those who worked in organizations that did not.

## PDCA

Figure 5:   PDCA Model applied to the Safety Culture Model

# 1. Safety leadership

Safety leadership is the practical manifestation of heroes in their daily lives by demonstrating a visible commitment to safety.

It can manifest in rituals: practice of behavioral observation, safety committee leadership, field presence, etc.

As for the symbols that demonstrate this element, we can have: messages and communications from the leadership.

Below, I have separated some questions regarding this element for your reflection:

☞ Do employees perceive their leaders as role models in safe behavior? Is the leadership present, abiding and exemplary?

☞ Since top and middle leaders live in safety, what visible commitment demonstrations do they perform on a daily basis?

☞ What are their safe habits?

☞ Does leadership really believe in safety? Does it agree with the safety rules and model adopted?

☞ Are they spokespeople for safety values and beliefs?

Check out the main aspects of safety leadership in each type of culture:

**Sustainable**

**5**

Leadership acts as a safety leader and is recognized as an example within the site. This is reflected on the results achieved at the unit (improved performance for 3 consecutive years). Performance in its area is an example on all KPIs. Senior leadership understands its role and importance as a spokesperson for safety culture, creating exposure opportunities through a genuine message to the entire organization. There are no more Safety goals. The leaders constantly challenge themselves to seek innovative ways to show their visible commitment to safety.

**Proactive**

**4**

Leadership understands the value of and empowers itself with safety, and begins to exercise its leadership by practicing See & Act. Safety performance is part of their business plan and is reviewed at strategic meetings. The emphasis of the leadership is to be an example, build a legacy, and act deeply by challenging their teams with goals, appropriately allocating funds and revitalizing safety efforts.

The accountability process happens and is effective with results and critical analysis, always seeking opportunities for improvement. Senior leadership understands its role and carries out visits without notice with the purpose of strengthening the safety culture at the site.

Leadership feels responsible when occurrences happen and thinks of the failures in their behavior that have allowed accidents to happen.

**Bureaucratic**

**3**

Leadership understands its role and responsibility in safety and receives as non-technical expertise to develop: safety leadership, with a defined routine, and there are achievement goals. Leaders seek to achieve goals to avoid embarrassment. The performance of their role, responsibilities, and routine depends on the safety team. Senior leadership begins to visit the site without a safety system, however a few non-structured safety messages are shared. The accountability process happens, but it is not effective and has little result. due to it being poorly structured. Leaders do not feel responsible for occurrences and always look for the guilty parties.

**Reactive**

**2**

Leadership manifests itself and leads in safety always after an occurrence and/or whenever there is a risk to business continuity. The organization's senior leadership routinely visits the site after a serious occurrence with losses and damages focused on looking for guilty parties. However, there is no formal accountability process, from the definition of roles and responsibility. Leadership reviews safety performance with the safety area.

**Pathological**

Leadership does not lead in safety and does not believe this is important and necessary. Leadership visits to the site do not happen with a focus on safety. There is no kind of accountability for safety. Safety goals are focused on reactive indicators.

To get to the sustainable level, see the **ALWAYS** and **NEVER** recommended.

---

## ALWAYS

- Responsible for the team's safety;
- Example, reliable and upstanding;
- Acts in compliance with safety procedures, standards and rules;
- Understands and proactively acts in the construction and strengthening of the safety culture;
- Fosters innovative thinking;
- Genuine in the practice of safety values;
- Inspires, encourages, and engages people towards a safety attitude;
- Practices empathy, is ethical and fair;
- Tirelessly communicates safety;
- Practice Seeing & Acting;
- Present on the field practicing safety, walking the way he talks;
- Includes safety in strategic meetings and in the unit's strategic plan;
- Demonstrates visible commitment to safety beyond established goals;
- Routinely acts with visible actions in favor of prevention; and
- Has a personal safety message.

---

## NEVER

- Be reactive and punitive;
- Have a police posture;
- Be underhanded in the practice of safety;
- Procrastinate unsafe situations;
- Delegate your role and responsibility;
- Say and do things you don't believe about safety;
- Be inconsistent;
- Fail to comply with your safety commitments and routine;
- Believe that safety is the responsibility of safety alone; and
- Fail to comply with safety rituals.

## 2. Safety Policy, Principles and Values

This element represents the safety governance model, where it is positioned and how it is handled.

They can manifest in rituals: Safety committee, annual goals plan, investment plan, etc.

The Policy can be understood as the company's letter of commitment, translating into a formal document that must be widely communicated, showing the commitments assumed with all interested parties. The principles are the basis for sustaining the Policy, they are the set of guidelines, which are reflections of the beliefs. The values alone show what is not negotiated for the organization.

They can be manifested in the symbols: communication plan, structural conditions of the units, values, beliefs, and spokespersons.

Below, I have separated some questions regarding this element for your reflection:

☞ How does this set present itself within the company?

☞ What is the governance and organization of safety belief like?

☞ Is the commitment to safety established in company policies? Is protecting people's physical integrity a value?

☞ What are the safety principles that the organization has established?

**Sustainable**

**5**

Employees are passionate and believe in the company's safety principles and values. The company is recognized as world class for its demonstrations of visible commitment in establishing and fulfilling commitments made. It is desirable as a place to work and do business. Work safety is an unquestionable core value. It is understood that both business excellence and safety excellence are achieved with the same actions. The organization refuses to place other performance objectives above safety.

Check out the main aspects of policy, principles and safety values in each type of culture:

**Proactive**

**4**

The organization establishes safety as a value and revitalizes its policies, principles and values. Leadership can translate commitments undertaken with practical examples while interacting with its employees, and employees understand their role and responsibility.

Practices are consistent with that set forth on paper, and this is noticed within and outside of the organization. Safety performance is considered a priority for the organization.

**Bureaucratic**

**3**

The organization sets out its policy and principles with the purpose of showing visible commitment to safety; its priority is to implement the safety management system. These documents are known by the leadership and safety area. Decisions of the middle leadership do not match statements given by the company. Managers speak of operational safety, but do not always follow what has been said, particularly when cost and production goals are not reached. However, the organization realizes that having a policy, principles, and values may appeal to its workforce and customers, but its priority is always to meet safety goals.

**Reactive**

**2**

The organization drafts a safety policy with the purpose of setting out safety guidelines. However, this document is little known, its management is done by the safety team. In addition, leadership believes that the organization implicitly has the safety value in its value statement. Keeping operations on track is the number one priority. The documents are presented as a tool for dialogue with the government and customers. The organization does not see value in these documents and they do not reflect operating practices. The priority is to meet cost and productivity goals.

**Pathological**

**1**

The organization does not have a record or bond of safety in its policy, principles, and safety values.

Cost reduction and production are the only priorities. Safety is seen as a cost. The principle is to do things as cheaply as possible.

To get to the sustainable level, see the **ALWAYS** and **NEVER** recommended.

---

## ALWAYS

-  Safety commitments must be Top Down;
- Commitments must be cascaded to all employees and be consistent with what is established in policies, principles and values;
- Everyone must know their role and responsibility;
- Top and middle leadership must genuinely agree and commit;
- Maintain up-to-date and visible communication of safety policy, principles and values;
- Challenge everyone to commit and share responsively by going outside the unit's gates;
- Engage the market with its best practices in achieving challenging goals;
- Check the adherence of the execution and fulfillment of the commitments in the middle leadership;
- Build a values manifesto with the participation of all employees; and
- Senior leadership must be the spokesperson for safety policy, principles, and values.

---

## NEVER

- Imply the company's commitments and values regarding safety;
- Fail to formally commit to safety;
- Allow the safety area to unilaterally make safety commitments;
- Assume and define commitments that are not suitable for your scenario and impact;
- Allow these commitments to become obsolete and unreliable within the organization;
- Allow these commitments to be broken;
- Allow for negotiation in the achievement of goals in order to take credit for the process;
- Stop carrying out a critical analysis in the definition and fulfillment of the assumed commitments; and
- Think safety is the only area that should be accountable for safety in the unit.

# 3. Risk Perception and Management

Perceiving and managing risk is a skill for everyone in the organization, they must permeate all levels and explain our decision-making model.

They can manifest in rituals: gemba walk, risk analysis, investment plan, etc.

They can manifest in the symbols: rate of resolution of unsafe conditions, safety conditions of machines, equipment, etc.

Heroes have the role of supporting/participating and demonstrating their insight and commitment to risk management.

Below, I have separated some questions regarding this element for your reflection:

☞ How do people perceive risks and anticipate them?

☞ How is the decision made to manage from critical risk to low risk?

☞ Is there a hierarchy in risk control?

☞ Who participates?

☞ Do people feel they are heard when reporting a risky situation?

☞ Is the environment safe to talk about safety opportunities/vulnerabilities in the unit? or can punishments occur for those who show concern for the safety of the unit and the surrounding community?

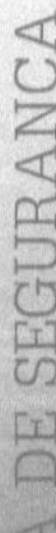

**Sustainable**

**5**

People empower themselves with risk management and the "What if?" failure thinking at the unit and expand the concept to activities outside of work. The process is constantly recycled and fed back within the organization.

Everyone feels bothered and constantly challenges themselves: is there something we aren't seeing and treating? Do we have a false sense of safety?

**Proactive**

**4**

There is a risk management system focused on all areas of the organization, the entire senior and middle leaderships know and use the control hierarchy in their see & act and prioritizations. The "What if?" is used in organizational dialogues. Employees can use risk management in their routine. Failure and variability thinking are strengthened throughout the organization. Analysis results are disclosed with the concern of ensuring people's understanding. Each employee understands their role and responsibility in risk management and the use of tools. Employees are empowered to notice and manage risks proactively.

**Bureaucratic**

**3**

There is a structured Risk Management system focused on the production area, however there is no clarity regarding the criteria of the methodology used, and the whole management, communication, and execution depend on the Safety department. The operational front line is not involved in the Risk Management process. However, there is no clarity regarding plans and executions. Everything is documented and is with the safety department to be submitted during audits.

**Reactive**

**2**

Risk management is done with an emphasis on legal requirements and post occurrence. The process is conducted by the safety team. There is no structured risk management and perception system. People who notice risks and speak up are punished.

The priority is to not stop the unit; actions are taken always with a focus on critical equipment that can compromise their goals.

**Pathological**

**1**

There is a focus on killer risks that prevent the functioning and/or stall the unit. The environment is not safe for talking about safety, people who do so are not seen in a good light within the organization.

To get to the sustainable level, see the **ALWAYS** and **NEVER** recommended.

## ALWAYS

- Leverage the "What If?" failure thinking;
- Everyone must know the risks and the control hierarchy;
- Ensure that the risk analysis is prepared by a multidisciplinary team;
- Involve the activity's employee in the entire risk perception and management process;
- Use clear methodologies and simplicity in the details;
- Empower everyone in the See & Act practice;
- Check adherence to understanding of associated risks;
- Create diverse learning experiences in order to enhance knowledge of hazards and risks for safety acculturation;
- The Area Owner must be responsible for analyzing, updating, and disseminating the results and risk management control measures;
- Use risk analysis in a playful way in integration;
- Listen to operations so that risk analysis are updated considering the variability of the business;
- Be uncomfortable and generate discomfort in employees when they stop identifying risks; and
- Build the Risk Perception and Management skill across the organization, regardless of hierarchical level.

## NEVER

- The safety area must carry out the entire risk analysis and management process;
- Fail to establish prioritization criteria;
- Delegate the non-delegable, allowing third parties to perform risk analysis with off-the-shelf solutions;
- Hide the results of the risk analysis;
- Ignore possible failures and variability;
- Restrict and treat the person who thinks about possible failures as negative;
- Allow an employee to start their activities without knowing the dangers and risks;
- Perform the dissemination of analyzes only by murals or matrices;
- Ignore the development of this skill for all roles;
- Fail to give feedback regarding perceived risk;
- Fail to take action in the face of an identified risk situation; and
- Believe that your work has no risk and do not allow this belief to any employee.

# 4. Challenging Plans and Goals

This element brings direction to safety practices, establishing a plan for achieving goals. The plan and goals must be part of the safety committee's agenda and must be widely discussed so that they are challenging and achievable.

How it is manifested in rituals: Safety committee, critical safety analysis, strategic safety plan, among others.

How it is manifested in symbols: Tables and reports etc.

Below, I have separated some questions regarding this element for your reflection:

- What are the safety goals? Are they attainable, are they challenging?

- Were they just based on reactive indicators or are they a mix of reactive and proactive ones?

- Are they prepared within the SMART model?

SMART (S — Specific; M — Measurable; A — Attainable; R — Relevant; and T — Time-Bound)?

Check out the main aspects of plans and challenging goals in each type of culture:

**Sustainable**

**5**

The committee, with the participation of senior leadership, sets out and manages goals. The main emphasis is on building a zero-occurrence environment. Senior and middle leaderships encourage all employees to seek and build this result. Each employee knows their role and responsibility for its achievement. Additionally, there is as aspiration to build a safety reputation for the brand.

**Proactive**

**4**

The goal plan is set by the committee, who uses commitments undertaken in the policy, principles, and safety values as a basis, in addition to bringing the organization's updated risk analysis and the management system's elements. The focus is on the quality of the fulfillment of each goal. Indicators are disclosed according to the management practices in sight. Failure to meet the goals impacts recognition actions. Goals are achievable and challenging, reflecting the search for strengthening the culture; all employees know the goal plan and their respective role and responsibility.

**Bureaucratic**

**3**

The organization sets short- and medium-term plans and challenging goals, still in an isolated manner not considering the organization's business plan. The goals take into account legal conformity and the organization's risk analysis. Proactive indicators, with an emphasis on the amount of reports, are established. Generating little learning for the organization ("we have gone xxx days without an accident with leave"). Indicators are disclosed only at results meetings. They are managed by the safety area. Failure to meet the goals does not impact any recognition action, as safety KPIs are not tied to them. They are also not tied to the leaders' career path.

**Reactive**

**2**

Reactive indicators are set out with the purpose of meeting legal requirements and ensuring business continuity, however such indicators are managed by the safety area. A critical performance analysis is not conducted, and there is no accountability over safety results.

**Pathological**

**1**

There are no performance indicators and management for reactive and proactive safety KPIs (key performance indicators)

To get to the sustainable level, see the **ALWAYS** and **NEVER** recommended.

---

## ALWAYS

- Establish safety goals that speak to all areas and must be aligned with the strategic planning of the business;
- The goal plan should reflect reactive and proactive KPIs;
- The discussion, establishment and management of the goal plan must be done by the safety committee;
- The goals must reflect the commitments made in the safety policy, principles and values;
- Targets should consider the results and needs of the risk analysis;
- Establish a learning process and critical analysis of goals;
- Achieving goals must be linked to a Reward, Recognition, and career path program;
- Ensure that all employees are aware of the goals and their respective Role & Responsibility;
- Goals must be set in SMART format;
- Establish massive communication to ensure that all employees are aware of the company's goals and objectives;
- Senior leadership needs to treat safety goals as carefully as they do financial metrics; and
- Senior leadership must be an ambassador for the goals and engage employees in the search for construction and achievement.

---

## NEVER

- Establish safety goals only for the safety team;
- Work only with reactive goals;
- Set goals without criteria and unattainable;
- Fail to communicate and update status;
- Lack transparency in the goal management process;
- Forget the purpose of safety goals, which reflects the quest to prevent people from getting hurt; and
- Allow the achievement of goals to be linked only to the quantity of delivery, but also the quality of deliveries.

## 5. Establish Priorities and Allocate Resources

Priorities and the allocation of any type of resource support a safety culture. That is why it's important that it be discussed and that it considers operational, tactical, and strategic variables.

They can manifest in the rituals of: Safety committee, risk analysis, audits, etc.

They can manifest through the symbols: Structures that demonstrate allocated resources, communication practices, etc.

Heroes are the protagonists in approving plans and allocating resources.

Below, I have separated some questions regarding this element for your reflection:

- Does the company have a matrix of priority investments?

- What are the criteria that make it a priority?

- Is it legal compliance or risk management? Do people know the safety priorities and understand why they were placed in this hierarchy?

- Is there transparency when resources are allocated?

- Is there a methodology for decision making that considers legal variables, exposure to risk and other criteria deemed necessary?

- Do employees know this matrix?

**Sustainable**

There is a plan with a budget for innovation and optimization, adding value and leveraging business performance. The risk analysis is updated, and new investments are made as needed. It is noticed that investing in safety contributes to leverage the organization's other financial results.

**5**

**Proactive**

The organization has a committee prepared to understand and set safety investments, considering the risk analyses and management tools, using control hierarchy to make decisions. The investment plan is tied to the organization's strategic safety planning, with short-, medium- and long-term investment objectives and goals.

Funds allocated to safety are protected and do not run the risk of being redirected.

**4**

**Bureaucratic**

A safety investment plan is carried out, but without analysis structuring and prioritizing. The safety area begins to manage the investment plan with the unit leadership. Fund allocation is done to eliminate unsafe conditions and prevent accidents, but funds may be redirected, "whoever yells the most gets the most."

**3**

**Reactive**

Saving money by cutting costs is important, but the money is spent so that the Safety improvement is in line with legal requirements, responding to complaints and audits of regulatory agencies. The continuity of operations is the number one priority. Safety funds compete with maintenance funds.

**2**

**Pathological**

There is no safety investment plan. Investments are occasional and make up a maintenance portfolio.

Safety is seen as something expensive, and the only important issue is to prevent extra costs and business stoppage.

**1**

To get to the sustainable level, see the **ALWAYS** and **NEVER** recommended.

---

## ALWAYS

- Have a strategic safety plan that addresses risk analysis, legal compliance issues, reports of unsafe conditions and other tools that support safety risk management;
- Keep risk analyzes reliable and up to date;
- Investments must be accessed after the change management has been carried out;
- Have a safety investment portfolio;
- Make decisions using the control hierarchy;
- Ensure multidisciplinary discussion and decision;
- Seek optimization and innovation in hazard and risk mitigation projects, adding value to the business;
- Have skilled people involved in the decision-making process;
- Involve the areas of: safety, quality, maintenance, operator, area leadership and sanitation in the change management process to allocate resources; and
- Communicate the reason for prioritization and allocation of resources to those involved.

---

## NEVER

- Reactively allocate resources;
- Work the allocation of resources isolated from other areas;
- Leave the resource allocation criteria subjective;
- Procrastinate;
- Rationalize the danger and risk due to lack of history of occurrences;
- Think in isolation with "safety capex" thinking; and
- Allocate resources just because you will be audited.

# 6. Safety Team

The safety team needs to understand their role as architects of building and strengthening the safety culture. As we mature, we experience each stage of culture and our role is consolidated as one of great influence, stimulus, and inspiration for the organization.

We act directly and indirectly in all rituals; symbols and we can be considered as the neck of the heroes showing the direction with proactive stimuli.

Below, I have separated some questions regarding this element for your reflection:

- How is my safety team? A police type, imposing fear, reactive? Or a support team?

- Do employees want to be part of the safety team? Is working in the safety department part of the career path?

- What kind of influence does the team exert within the organization?

- How is the safety team perceived within the organization?

SAFETY CULTURE

**Sustainable — 5**

The standard of these professionals is distinguished, even by the type of activity they perform within the company. The focus is on supporting customers and their demands in the search for maintaining operational excellence in safety. The organizational safety habits are established and safety responsibilities distributed throughout the company, and the importance status of this department can be compared to the others. The statistic safety data have the same weight as the organization's financial performance indicators. This team seeks technologies and innovations to mitigate risks and offer a distinguished technical support.
This team treats safety in a humane manner and inspires the entire organization.

**Proactive — 4**

There is empowerment of all areas. We have the so-called Safety Owners. The safety team offers technical support to all areas. The support is focused on the challenge of maintaining and inspiring safe practices, seeking innovative solutions, looking at the root cause. All people are aware of the purpose of the safety area. The safety manager reports directly to the unit manager. There is a career path and competence matrix for professionals in the area.

In addition, for leadership and/or high potential positions, performing some activity in the Safety Department for some time is part of the development career path.

**Bureaucratic — 3**

Safety rituals and practices are implemented and led by the safety team. There is little maturity of other areas and there is strong reliance on the safety department. Empowerment is the major challenge and objective at this stage. The safety department has statistical analysis. The safety area reports to HR and/or to the quality and/or operations department. There is no training and competence building program for the safety team.

The perception is that there are too many bureaucratic activities. Finding the safety team on the field is sheer luck.

**Reactive — 2**

In Brazil, we can have the SESMT - Services Specialized in Safety Engineering and Occupational Medicine serving as technical support. The monitoring of the safety team is distant and works with punctual demands, focused on compliance with the law and treating occurrences, much like a "firefighter" within the organization.

Professionals are seen as little proactive, and being part of this department is being at the end of one's career.

**Pathological — 1**

There are no safety professionals at the unit. The area is not deemed necessary. Anyone can do safety.

To get to the sustainable level, see the **ALWAYS** and **NEVER** recommended.

---

## ALWAYS

- The safety team must support doing things "with" safety;
- The safety team must have clarity in its role and responsibility;
- The safety team must advance in safety technical support seeking innovative solutions and challenging current management;
- The safety team must be present in the field and known in the organization;
- Have a reputation and exert influence throughout the organization;
- Plan routines with an emphasis on providing the best customer service;
- Have inspiring leadership;
- Bring on the difficult conversations and look for synergistic solutions across all areas;
- Practice endomarketing with clear and impactful actions;
- Have a partnership and open attitude with clear principles and values "not negotiating the non-negotiable"; and
- The safety team must receive training and have its skill developed as a support area.

---

## NEVER

- Be reactive in offering technical support;
- Be recognized for the "can't";
- Be accommodating and focus only on complying with legislation;
- Present a discredited approach to the importance of safety;
- Be seen as a policeman exercising little power;
- Assume the precautionary principle without seeking technical depth; and
- Stay in the comfort zone.

# 7. Safety Committee

The safety committee is the main safety discussion forum. It is empowered, trusted, and led by members of upper and middle management.

The Committee is a ritual of manifestation of the safety culture and feeds other rituals and symbols through its performance.

The participation of the heroes in the committee is of fundamental importance for the ritual to be complete.

Below, I have separated some questions regarding this element for your reflection:

- How does the organization experience safety in an integral way?

- What is the safety adherence at all levels of the corporation?

- How is action taken safely? And the empowerment of areas?

- What are the forums that safety is discussed within the organization?

- How is safety decision making?

- What levels and areas participate in discussions and decision-making?

- What is the safety committee's accountability process like?

## Sustainable

The committee begins to conduct the unit's strategic safety planning. Subjects are 360°. Discussions evolve from proactive indicators. There is a strengthening of the safety reputation due to demonstrations of visible commitment of the committee, of which the safety team is a member.

The committee challenges the organization to reach and maintain excellence in safety.

**5**

## Proactive

The committee is now managed by different areas on a rotating basis. The actions it discusses are not only regarding the unit but also its surroundings, with the involvement of stakeholders. The KPIs discussed are reactive and proactive. Participants are middle leadership leaders of all areas of the unit. Additionally, a mirror committee is established with members of senior leadership with a specific agenda, strategy and tactics regarding the strengthening of the company's safety culture.

The safety area technically supports the discussions.

The committee is recognized by employees and has strong credibility within the organization.

**4**

## Bureaucratic

The safety committee is established with a work plan and agenda, but it is managed by the unit's safety team. KPIs monitored are at the top of the pyramid. Topics of internal interest to the factory are discussed. Participants are from the operational/industrial area.

The committee is not known by employees and does not have practical performance. Meetings are long, and little benefit is perceived,

**3**

## Reactive

Meetings among the unit leadership occur only after an accident with losses and damages. The entire process is coordinated by the safety team. Any interest diminishes when things start improving/normalizing.

Decisions related to the safety agenda happen between leaderships and the safety team.

**2**

## Pathological

There is no senior and middle leadership committee and no structured meetings are held to strengthen the Safety Culture and discuss actions.

**1**

To get to the sustainable level, see the **ALWAYS** and **NEVER** recommended.

## ALWAYS

- Build a reputation;
- Have specific actions for committee members;
- Count on the participation of the organization/unit leaders;
- Have a defined agenda;
- Have meetings that meet the defined time;
- Be a decision-making forum;
- Focus on strategic safety planning;
- Take a systemic approach;
- Listen to the organization;
- Receive feedback on your performance;
- Report to the unit for the work carried out;
- Have a disciplinary matrix for members to manage your absences; and
- Conduct a safety performance review of the unit.

## NEVER

- Be led by the safety team;
- Be an embarrassing, reactive and attention-grabbing forum;
- Discuss only reactive indicators;
- Procrastinate decision making;
- Stop sharing/communicating decision making;
- Keep rescheduling meetings;
- Discuss only specific topics;
- Allow a single leader to lead the committee;
- Lose the credibility of the committee; and
- Stop having a strategic and tactical committee in the company.

# 8.  Tools that save lives

It's an important and strengthening element of the safety culture, which brings together the entire tooling of checklists, inspections, reports, risk analysis and permissions used by the organization.

It's responsible for several safety rituals, such as: work permit, inspection of machines and equipment, among others.

These rituals materialize symbols, which can be identified, such as: Loto - Lockout & Tagout, PPEs, safe conditions, among others.

Heroes have, in their work routine, the use of various tools that save lives.

Below, I have separated some questions regarding this element for your reflection:

☞ Are lifesaving tools known, trusted, and used throughout the organization?

☞ Do our employees know the purpose of lifesaving tools?

☞ Is adherence and quality of use of lifesaving tools analyzed?

☞ Does the organization remain uncomfortable and asks itself: What might our lifesaving tools not be seeing?

Check out the
main aspects
of life-saving
tools in each
type of culture:

**5 — Sustainable**

There is an empowerment of life-saving tools. Most tools are concentrated on the activity planning steps. There is a complete planning process with anticipation of problems and revision of processes. There is less paper, more thought, and the planning process is well known and discussed. Processes are improved by employees themselves, the use of technologies that improve the reliability level and risk perception is enhanced.

**4 — Proactive**

Tools are now prepared by the area owners, with the involvement of the safety area as support. People perceive tools as life-saving, always done on the field, according to the procedure. The purpose is to achieve the results/efficiency that the tool can yield.

Safety tools are integrated into the unit planning. Leadership realizes that its participation in and involvement with the tools are a layer of control to prevent accidents, and thus makes sure to demonstrate visible commitment by using the tools.

**3 — Bureaucratic**

A risk analysis is conducted, and permit/clearance systems and other safety management tools such as checklists and inspections are established. The process depends on the safety team. The approvers and people involved do not see value and thus many times they are done incoherently and do not have credibility within the organization.

The number of reports is used to show that the system is working, with an emphasis on the amount of tools used. There is no qualitative tool analysis. When accidents happen, they wonder: Wow... But we have so many tools and do so much for safety...

**2 — Reactive**

Preventive tools are only implemented for activities recommended by law. The work safety area is responsible for the process.

There is little systemic use of such tools, and communication is not adequate. The perception is that it is delaying the company's processes.

**1 — Pathological**

There are no tools that manage critical works, and the perception is that it is not needed. Work planning is focused on the fastest and cheapest completion of the work.

"The belief practiced among employees is Take care of yourself."

To get to the sustainable level, see the **ALWAYS** and **NEVER** recommended.

## ALWAYS

- Look for simple and deep tools that prevent accidents;
- Critically evaluate the performance of the tools and seek improvements;
- Seek leadership engagement and involvement in the use and performance improvement of the tools;
- Be appropriate to the business (activity profile);
- Have reactive and proactive indicators on the performance of the tools;
- Assess the quality of training and use of tools;
- Check the adherence and understanding of the tools in the field;
- Perform benchmarking;
- Listen to all the areas involved to ensure the credibility of the tools;
- Use the pilot tools format before implementing a substantial change; and
- Focus most of the tools on the job planning stages. Understanding that the act of planning and executing as planned constitutes a control layer.

## NEVER

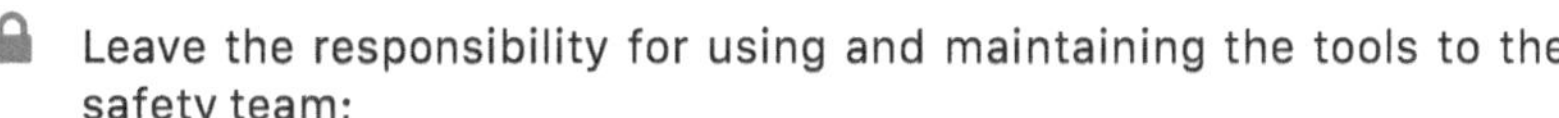

- Leave the responsibility for using and maintaining the tools to the safety team;
- Bureaucratize the development of activities;
- Show more importance to paper than to people;
- Look for obsolete solutions;
- Focus on quantity rather than quality;
- Choose tools that are impossible to use; and
- Lack understanding of the importance, reason, and value of existence of each of the tools.

# 9. Rules that save lives

The rules that Save Lives represent the limits of care for our employees. They must be established from the history of more severe occurrences/accidents and present eminent risks. They must be linked to a program of progressive motivation, as their non-compliance can generate losses and damages of the most severe level and in this regard, non-compliance must have consequences.

All rules must be possible to be complied with and their dissemination must be broad, massive, and constant.

They are considered symbols of care at its highest level.

Below, I have separated some questions regarding this element for your reflection:

- What are the non-negotiable rules within the organization when it comes to safety?

- Are they clear?

- Are they tied to a model of motivation for life?

- Do employees understand the purpose of the rules?

- Is it clear where they came from and why they were chosen?

- Are life-saving rules for all employees, third parties and visitors?

Check out the
main aspects
of life-saving
rules in each
type of culture:

## Sustainable — 5

Employees observe the Life-Saving Rules and understand the importance of the Motivation for Life Program. Leaders have frequent interactions with their team to test their understanding of Life-Saving Rules. In addition, there is an assessment of the nature of violations/investigations tied to Life-Saving Rules to address behavioral issues.

## Proactive — 4

Leadership feels that it is an integral part of the management process, monitoring compliance with the golden rules and applying the relevant disciplinary measures. Rules undergo a process of change and focus of its understanding. They are humanized and are now referred to as Life-Saving Rules, comprising an act of care performed by the organization, setting the safe limits for every person within the unit. These rules remain tied to a motivation for life process and are based on catastrophic (serious at the Unit) accidents and critical risks and are widely publicized.

## Bureaucratic — 3

A study is conducted, and golden rules are established. A communication process is carried out with a punitive focus. Leadership does not feel responsible for ensuring compliance with these rules. The rules are perceived as safety rules. They are mostly applied to third parties.

## Reactive — 2

A few rules are communicated on an informal and occasional basis, usually after an accident. No analysis process is conducted. Employees recognize the Company's Code of Conduct as a rule.

## Pathological — 1

There are no formally established safety rules.

To get to the sustainable level, see the **ALWAYS** and **NEVER** recommended.

## ALWAYS

- Build rules from the unit's history of catastrophic accidents and killer risks;
- Ensure proper communication and understanding at all levels of the organization;
- Implementation and review must precede an adaptation period;
- Check the understanding of the rules across the organization;
- Check adherence to the rules;
- Apply the rules fairly and equally;
- Critically analyze and define hierarchies between rule and wrongful act;
- Act immediately upon non-compliance; and
- Set review period for life-saving rules.

## NEVER

- Build rules from punctual occurrences;
- Imply the rules;
- Establish rules that cannot be followed;
- Establish different consequences for each rule; and
- Negotiate breaking the rules.

# 10.  Motivation for life and duty to refuse

Motivation for life is a progressive ritual, it is a learning process, where we recognize and reward employees and third parties who comply and stand out with their safe behavior and also associated with this same element we apply consequences for those who insist on breaking the rules and commit wrongful acts.

The symbols that materialize the motivation for life are artifacts arising from recognition and reward, and also what evidence the consequence of breaking the rules.

There is a routine that involves the heroes in the entire ritual of motivation for life, starting in the design processes of this program and being realized in the acts of recognition, reward, and the application of disciplinary sanctions.

Regarding the duty of refusal, it is a reactive tool, where the employee needs to trust his leadership and the care practiced and perceived. We must strengthen safety so that this tool becomes obsolete as we mature in our safety culture.

I have separated some questions below to reflect on the use of motivation for life and duty of refusal:

☞ What, who and why do we recognize when it comes to preserving lives? Motivation, it's important to say, works within a process of discipline, it is not limited to recognizing who acts correctly. Giving yellow and red cards for unsafe acts is just as important in motivation as congratulating.

☞ Is the process perceived as fair and equal?

Check out the main aspects of motivation for life and duty of refusal in each type of culture:

## Sustainable — 5

The organization understands the importance of the Motivation for Life program and applies it with principles of justice and equality. The Duty of Refusal becomes obsolete.

It is understood that acknowledging the good performance in Safety is very valuable and motivates people without having to reward them in other manners. The acknowledgement is feeding back to the career path within the organization.

## Proactive — 4

The organization has a committee and investigates all rule breaking occurrences reported. The understanding is that disciplinary measures are not punishment, but care. The duty of refusal is used and understood as a life-saving tool. Good Safety performance is acknowledged in reviews for promotion. Employee reviews are based on the execution of correct proceedings and the occurrence (or lack thereof) of incidents. Reward practices are used sometimes, but this is not prioritized.

## Bureaucratic — 3

The Progressive Motivation Program is cascaded down and communicated. A committee is established to investigate and handle breaks. Deviations are investigated, but there are still difficulties in applying the consequence management with principles of justice and equality. The Duty of Refusal is a feared tool. The focus of application of the consequence management is on third parties. It is commented that good performance in Safety is very important. Awards are given out in the name of safety. There are competitions and games to recognize safety awareness. Accident rates are used to calculate bonuses.

## Reactive — 2

The Program exists, but it works with a few formalized warnings. It still does not occur in an organized manner, it is a response to occurrences that have caused severe losses and damages. Employees that use the Duty of Refusal are penalized. Offering rewards for good performance is not common. Bonuses are reduced when accidents happen.

## Pathological — 1

There are no formal safety rules and guidelines and no discipline/consequence management model, the Duty of Refusal is not recognized. Practicing acknowledgement is also not seen in good light, as staying alive is already a great reward.

To get to the sustainable level, see the **ALWAYS** and **NEVER** recommended.

## ALWAYS

- Have a spokesperson who is an important company executive;
- Use principles of fairness and equality to permeate all levels of the organization;
- Use failures as learning opportunities;
- Have credibility;
- Form a committee within the unit to assess individual breakdowns and failures and conduct an investigation process;
- Define simple and clear criteria;
- Ensure communication at all levels of the organization;
- Assess the responsibility of the immediate leader;
- Strengthen the practice of the Duty of Refusal, it is an act of care for the preservation of life;
- Recognize employees who make a difference in safety;
- Recognition programs must be short term/short duration;
- Recognition programs must be linked to reactive and proactive performance, built on SMART KPIs;
- Offer, as a reward for recognition, moments of enjoyment with family and friends; and
- Senior leadership must participate in recognition actions: certificate delivery, lunch, presentations, etc.

## NEVER

- Embarrass those involved;
- Treat as a punishment;
- Be reactive;
- Analyze the facts unilaterally;
- Lack transparency;
- Jump to conclusions;
- Stop listening to those impacted;
- Assess the liability of only the employee who broke the rules;
- Fail to include human factors within the analysis;
- Fail to apply disciplinary measures to senior leadership;
- Implement the recognition program using reactive indicators as criteria;
- Allow the reconnaissance program to be an action isolated from the safety area; none of that! It needs to have relevance within the organization;
- Believe only in money. Awards, no matter how simple, need to be sincere and genuine;
- Work only on individual recognition and rewards, involve the team in order to develop a sense of team; and
- Leave the recognition program criteria subjective.

# 11. Management of Contractors and Visitors

Contractors and visitors need to be influenced by the culture experienced in our company so that they can understand how their safe behavior can contribute to the entire organization. Both groups perceive the manifestation of the safety culture, through the rituals, symbols, attitudes, and behaviors of the heroes (contractors) and the existing communication practices.

I have separated some questions below to reflect on the management of these groups within the organization, we understand that we cannot have differences in dealings with contractors and visitors.

- What are the precautions for our contractors and who visits the organization?

- Can I pass on safe habits to these people?

- Is there a progressive development for permanent contractors?

- Is there a development, discussion, and accountability forum for contractors?

- Do visitors and contractors participate in our rituals and are they able to perceive the Safety Culture symbols?

- Do our heroes include targeting Contractors and Visitors in their messages and actions?

**Sustainable**

**5**

The contracted company and contracting areas understand the value of the contractor management process. The safety level and management are not negotiated for contractors. Work is allowed to be postponed until requirements related to the safety system have been met. Contractor management processes with the best performance are incorporated into a recognition program. Visitors receive integration, their adherence is checked and they are escorted within the organization and suggestions are requested on how we can improve safety during visits.

**Proactive**

**4**

The organization has an integrated program between safety, supplies and other contracting areas. Each one understands its role and responsibility. The program is linked to the motivation for life program. The contracted company and contracting areas understand the priority of the contractor management process. Pre-qualification of the contractors requires evidence that there is a safety management system in operation. The company and contracted companies make joint efforts in matters related to safety, and the company assists in training contractors. Visitors receive integration, their adherence is checked and they are escorted within the organization.

**Bureaucratic**

**3**

The organization has a Contractor Management Program including permanent and temporary. The program is managed by the safety area with little interface and liability of the contracting area. The program is seen as bureaucratic. Third parties must meet legal and pre-qualification requirements, based on questionnaires, indicators, and documentations. Safety standards go down when no third-party company can meet those requirements. Third parties must strive to live up to the standards through their own means. Visitors receive guidance and are escorted within the unit.

**Reactive**

**2**

The management of the contractor safety system becomes important after an accident. The most important issue when selecting a service provider is the price with minimum compliance with legal requirements, despite unsatisfactory safety performance not bringing consequences in the selection of third-party companies. No safety practices are established for visitors, the obligation is that the receiving area should provide safety information.

**Pathological**

**1**

There are no contractor and visitor safety controls at the unit. Contractors are expected to conduct their tasks with scope compliance and lowest cost. Safety-related problems are the full responsibility of the contracted company. No safety practices for visitors.

To get to the sustainable level, see the **ALWAYS** and **NEVER** recommended.

## ALWAYS

- The contractor management program must have clarity on the role and responsibility of the contracting area of the purchasing department and of the safety area;
- The contractor management program must comply with all the steps of the PDCA and have its critical analysis carried out periodically;
- Make sure everyone knows the contractor management program;
- Share lessons learned from this program;
- Positively influence contractors and visitors in the adoption of safety practices;
- Listen to suggestions and recommendations from contractors and visitors;
- Ensure proper communication with contractors at all stages of the program; and
- Work on defining a scope that considers the risks associated with the work and the variability that may occur.

## NEVER

- Have the management of contractors and visitors perceived as bureaucratic and reactive;
- Leave the management only in the hands of the safety area;
- Allow uneducated and unaccompanied visitors;
- Fail to comply with the program;
- Embarrass the contractors;
- Be unfair in the application of the program and its respective criteria;
- Allow subjectivity in the program criteria;
- Define contractors only using the cost criterion; and
- Leave the progressive motivation program open, without rules, roles, responsibilities, and penalties.

# 12. People Change Management

Often the focus is on managing changes in areas, machines, equipment, products, and tools, without seeing the employee within this process. How does the organization take care of people in these transitions? Are the skills it needs to develop to keep itself and others safe, in fact, developed?

Change management is a ritual and should not only permeate physical and material structures, we must consider people during all stages. As symbols that materialize care for people in change processes, we can highlight the ergonomic conditions, the training certificates that were offered. Leaders (heroes) must be those who drive this process to occur correctly and fulfill their role.

I have separated some questions below to reflect on how we experience change management:

👉 What do we understand as applicable to the change management ritual?

👉 Who can open a change management process?

👉 What are the areas that are involved in change management?

**Sustainable**

**5**

The program is known by all employees, everyone can initiate a change, a critical analysis is carried out periodically. The monitoring of changes is done at the safety committee. Adherence to and the benefit of the people management of change is visible within the organization.

**Proactive**

**4**

The program is updated, and the gaze and treatments are not only to structural and technological changes, it now considers change in people formally. Senior leadership sponsors this tool. This program is linked to the career path of the license to operate element. The program is managed by the change owner and is supported by the safety committee. The management of change is a competence, which is an integral part of the license to operate element for all employees.

**Bureaucratic**

**3**

The Management of Change Program is cascaded down for all kinds of changes. Roles and responsibilities are established. The program is managed by the engineering area with some interface from the safety and human resources departments. The emphasis of the management of change program is still on structural and technological changes, employees are not considered within the process and thus much rework is generated and projects usually exceed the original budget.

**Reactive**

**2**

There is no structured management of change program. A few actions are taken in major changes. Those actions are managed by the engineering area and are not linked to the license to operate matrix.

The safety area is involved after the change is implemented with the purpose of adapting it to the applicable legal requirements.

**Pathological**

**1**

There is no management of change program. This isn't even considered within the organization.

To get to the sustainable level, see the **ALWAYS** and **NEVER** recommended.

---

## ALWAYS

- Establish a program of changes that consider structure, physical conditions, technology and people;
- Share lessons learned from the change management process;
- Ensure training so that people can lead and adhere to the change management program;
- Link the change management program to the license to operate and career path program;
- Make the change management program accessible to all employees;
- Change management must be conducted by a multidisciplinary team;
- The safety area must technically support all changes;
- Ensure massive communication about the program; and
- Ensure simplicity and agility in the program.

---

## NEVER

- Disregard the changes;
- Rationalize the dangers and risks arising from the changes;
- Think that change management should be done by the safety area as an "entity";
- Establish criteria for change management based on reactive indicators.
- Use the change management program only for changes in physical, structural, and technological conditions;
- Allow undefined roles and responsibilities; and
- Allow a unilateral management program, that is, considering only the safety aspect.

# 13. Employee engagement

A sure way to shoot yourself in the foot is to ignore what's going on in our employees' heads. Safety culture is personal and when employees open up to new ways of seeing and understanding safety, it is no longer possible to return to a previous state where there was no such perspective. If we take the example of the transformation of a caterpillar into a butterfly, they are no longer able to revert to their previous physical forms.

How can we engage our employees in the real value and sense of safety?

Our efforts need to be in identifying the mindsets and/or limiting beliefs to reformat and/or re-signify them properly so that safety is not recognized by: Procedures,

Rules, PPE's etc. Safety means care. I need to take care of myself. I need to take care of the other. I need to allow myself to be taken care of by someone.

Additionally, understand that the mission of safety is to return home every day complete (with regard to physical integrity), well and at the agreed time.

Below, I have separated some questions regarding this element for your reflection:

- Do employees understand what safety means? They understand that safety is not a norm, a procedure, much less rules, it is defined by the practice of care.
- Do employees understand the mission of safety and share learning outside the company?
- What are the prevailing limiting safety beliefs present in my organization?

**Sustainable**

**5**

The levels of commitment and care are very high on all levels. They are conducted by employees who show a passion for living according to their high personal safety standards. Someone getting hurt is seen as a family tragedy. Safety concepts and practices are already understood as a value and taken outside of the work environment (family and personal life). Families are involved in using the tools and deploying good safety practices. The objective of employees is to influence the community in their understanding of the safety value.

**Proactive**

**4**

The workforce feels proud of its performance in practicing safety as a value and wishes to do better. People take care of each other and the environment. The proactive tools are known and used by the entire work force and all contractors.

Safety concepts and practices are treated as a priority to managers, and there is an attempt of humanization with an emphasis on behavior.

There are still opportunities in the understanding of allowing oneself to receive care and that we can stimulate our surrounding community in practicing care.

**Bureaucratic**

**3**

Employees are involved in using safety tools, work groups are formed to systematize actions and strengthen the safety culture. Employees agree that all of this management is important, but they don't always completely practice what they preach. The perception of employees is that safety meetings or actions offer a limited interaction between middle leadership and the operation, only for meeting goals, with an exclusive emphasis on reinforcing what we must do in safety. They realize that looking safe is worth more than being safe.

**Reactive**

**2**

"Take care of yourself" is the rule. Public statements on taking care of colleagues are made right after accidents, both by management and the workforce. This emphasis disappears after a period of good safety performance. Care is conducted by the safety team and considered a mere formality, with an exclusive emphasis on complying with the law.

**Pathological**

**1**

"Who cares about safety? We don't even do our physical examinations." Workers take care of themselves individually and according to their good sense. Care is seen as a waste of time.

To get to the sustainable level,
see the **ALWAYS** and **NEVER**
recommended.

## ALWAYS

 Ensure two-way communication is simple and clear;

 Engagement and care actions must be sensitive and humanized, preceded by the practice of empathy;

 Give feedback and show the results of actions with their respective benefits;

 Promote the practice of active care as a safety link;

 Engage employees so they understand their role and responsibility and the value of doing it safely;

 Use communication channels to listen to employees and account for requests received;

 Check the adherence and participation of the workforce in the safety practice; and

Involve the family and influence good safety practices in the home and community.

## NEVER

 Allow engagement to be done only by the safety area;

 Perform engagement actions only after the occurrence of serious events;

 Use practices that constrain people;

 Penalize the person for exercising care with their superiors; and

 Lack credibility and transparency in all processes.

# 14.  Behavioral Observation

How to turn safety into a relationship? The organization observes the safe and unsafe behaviors of its employees, in order to guide and correct those who are insecure, in addition to promoting positive reinforcement for exemplary behavior.

Does everyone understand that behavioral observation is nothing more than a structured safety conversation? Do they understand that this conversation includes the practice of seeing and acting, as well as active care? Also, do they know that everyone can perform a behavioral observation and at all levels? Does everyone feel confident in the practice of care inside and outside the organization?

Behavioral observation constitutes a ritual that can also be seen and experienced as a tool to strengthen the safety culture.

Leaders must maintain this ritual, which must be made accessible to the entire company.

**Check out the main aspects of behavioral observation in each type of culture:**

**Sustainable**

**5**

All employees start performing behavioral observation. There are no goals. Everyone understands the benefits and purpose of behavioral observation as a practice of active care. Everyone is fit to take care and allow themselves to receive care.

**Proactive**

**4**

Informal leaders, accident prevention agents, and emergency brigade members receive behavioral observation training and are recognized as observers.

Goals evolve into quality and not quantity of observation. Everyone understands the purpose and benefits of this tool. The observation system is now crossed between the areas. There is a reporting and data stratification system for decision making. Quality analyses are conducted in order to understand the causes of behavioral deviations.

Employees begin to understand and value the tool as a practice of care and start using these concepts outside of the company.

**Bureaucratic**

**3**

Here begins the journey of systemic application of behavioral observation; behavioral programs have their starting point, still in a tone of priority, and are leveraged by the leadership and safety team. The leadership and safety team are trained. The purpose and benefits of this tool are not yet clear. There are quantity goals for carrying out behavioral observation. There is a reporting structure. However, the tool is seen as punitive, and observed employees still feel embarrassed during the observation. Most conversations are superficial, as the approach is still robotic.

**Reactive**

**2**

The minimum required is done to meet legal requirements; this does not include observing behavior.

The discussion on behavior is reactive and always happens after a serious occurrence.

**Pathological**

**1**

Behavioral Observation is not done; this tool is unknown.

Informal behavioral observations, when conducted, happen with a punitive focus.

To get to the sustainable level, see the **ALWAYS** and **NEVER** recommended.

---

## ALWAYS

 Leadership must sponsor this program;

 Perform quality analyzes in order to ensure understanding of the cause of the behavioral deviation;

 Stratify the data generated in the behavioral observation;

 Build a relationship of trust with the observed;

 Lead with empathy and in a humanized way, encouraging CARE;

 Lead the conversation from an observation stage;

 Make a commitment to the observed;

 Allow the employee to reflect on deviations and the correct way to perform the task;

 Hear the observed;

 Understand why they do not perform the task safely;

 Go with the observed to resolve the deviation;

 For continuous line work, the immediate leader must be communicated about the performance of the behavioral observation;

 Recognize a good practice when identified;

 Explain what behavioral observation is and the importance of this tool to strengthen the safety culture;

 Thank them for their time and attention;

 Be focused on observation and never divide their time with other subjects; and

Use a handshake and eye contact to ensure closeness.

---

## NEVER

 Immediately correct the observed;

 Constrain and/or impose;

Raise your voice;

Punish or enforce the consequence management program;

 Conduct observation and conversation without interrupting the Observed's activities;

 Mix safety orientation issues and productive topics, vacations, etc.;

 Allow the worker to continue performing the task after observation without correcting the deviation;

Stop connecting with the life value of the observed;

 Complete the form during behavioral observation;

Expose the observed to their leader and colleagues;

 Use your cell phone during observation and/or any distraction tool; and

Work only with observation quantity targets.

## 15. Audit

Audits are seen as learning rituals and must be unannounced, so that they comply with their full cycle of evaluation.

Heroes should be the protagonists during audits showing the practice of safety leadership and seeking to learn from external cases so that their learning and safety realities can expand.

Below, I have separated some questions regarding this element for your reflection:

- Does the organization take advantage of what an audit offers to learn what the market is doing safely?

- Does it use this information as self-criticism and an opportunity for constant improvement?

- Can audits be announced or unannounced?

- Are Audits welcome and do they serve as an opportunity for improvement?

## Sustainable — 5

Cultural aspects related to the safety system are integrated into the organization. Critical analyses are integrated and interface with the entire business; they are not only done in management systems, but also in behavioral aspects.

25% of the workforce are internal auditors; everyone knows their role and responds in an audit. Audits continue to happen without notice.

## Proactive — 4

The organization understands the purpose of audits, which occur without notice. The entire auditing process is managed by the unit owner. Results are assessed critically, lessons learned are cascaded down, and actions are implemented. Results are communicated to everyone.

10% of the workforce are internal auditors; everyone knows their role and responds in an audit.
Audits confirm the efficiency and effectiveness of the implemented management system, and are no longer the main tool for driving safety improvements. Critical analyses are integrated and interface with the entire business. Cultural aspects start being measured.

## Bureaucratic — 3

The organization has a management system with structured audits and trained internal auditors. The focus is on meeting goals, audits are announced, and there is great preparation so that everything occurs in an acceptable manner. Results are communicated to everyone.
Critical analysis is not carried out in a way that integrates safety in all facets of the business. The important thing is to "pass" the audit.

## Reactive — 2

Process managed by the safety area. Audits are announced, and great preparation occurs so that everything happens as planned.

There is forced compliance with legal requirements during safety inspections. Audits and certifications are done according to stakeholder demands. "The main objective is to pass the audit and/or obtain the certificate."

## Pathological — 1

The organization does not have a structured internal and external auditing process. This is not considered important.

To get to the sustainable level, see the **ALWAYS** and **NEVER** recommended.

## ALWAYS

 Train internal auditors from all areas in a way that reaches 25% of the total employees, as internal auditors, and propagates the sense of ownership and clarity of Roles & Responsibilities;

 Have a fluid communication process with the unit;

 Share lessons learned;

 Perform critical analysis of the audit process;

 Recycle the training of internal auditors with practical and theoretical training; and

 Operate daily as if they were going to be audited.

## NEVER

 Allow only announced audits;

 Allow internal auditors to be members of the safety department only;

 Accept the perception that an audit is pure bureaucracy;

 Allow it to be punitive; and

 Get ready with improvements that send the message: "We only do that, because we are going to have an audit."

# 16. Benchmarking

Benchmarking is a learning ritual, strengthening the maxim that says that we should not only learn from mistakes, we should learn from those who are doing it right and are moving towards excellence.

The heroes are protagonists in the conduct of this ritual.

Below I have separated some questions for us to reflect on this element and its dimension within organizations:

- Does the organization talk to the market about safety?

- How much does it learn from it?

- Is there a system of exchange with the market that goes beyond the reactive indicators of safety management?

Benchmarking



## Sustainable

**5**

The organization has the best benchmarking practices. The program is defined, and challenges are frequent. Practices encompass several sectors. KPIs are established, and results are managed.

All levels of the organization are involved in identifying action points for improvement.

There is comprehensive communication on the subject and its learning. The Company perceives high added value to the brand for its efforts in maintaining the best safety practices.

## Proactive

**4**

The program exists and is linked to strategic planning, still having conduction and sector participation as its strength, and has a set periodicity. Communication is current and engaging. KPIs are established, and results are managed. The organization tries to be the best in the industry, Lessons learned are cascaded down.

## Bureaucratic

**3**

There is a benchmarking program, but it is not structured; it works to meet senior leadership demand only, without ties to the strategic safety planning. There is little communication and few results. The focus is on current issues that can be measured objectively and summarized using numbers. Lessons learned are not generated, and comprehensive actions are not implemented.

## Reactive

**2**

There is a benchmarking program for finance and production. A few safety reports and numbers are compared internally.

## Pathological

**1**

There is no benchmarking practice. This process is not seen as necessary.

To get to the sustainable level, see the **ALWAYS** and **NEVER** recommended.

<table>
<tr><td>

## ALWAYS

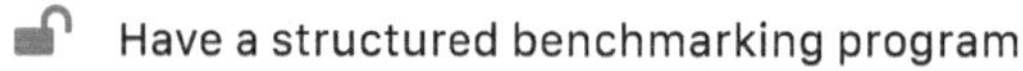 Have a structured benchmarking program;

Link this program to strategic planning;

Have KPIs that measure performance and adherence to shared benchmarking practices;

Participate by sharing and learning;

Ensure confidentiality of information;

Share lessons learned; and

Measure adherence to benchmarking practices.

</td></tr>
</table>

## NEVER

- Participate passively;
- Lack transparency;
- Work in an unstructured way, without being clear about what you are looking for;
- Fail to communicate the results of benchmarking practices;
- Stop benchmarking;
- Look for benchmarking, only from reactive themes;
- Stop learning and continually improve your processes; and
- Allow yourself to learn only from mistakes and practices that didn't work.

# 17. Occurrence Report

Reports are rituals in which everyone is a protagonist and can save a life.

It is necessary to signify the reporting of occurrences as an act of care.

Below, I have separated some questions regarding this element for your reflection:

- Are all occurrences really reported?

- Do employees understand the importance of reporting so that the episode is dealt with correctly and so that the repetition of these occurrences is avoided?

- Are employees engaged and confident in the process?

- The understanding of the report is that we are going to deal with the situation, or are we going to look for a culprit?

- Is our work environment safe to work in and to talk (report)?

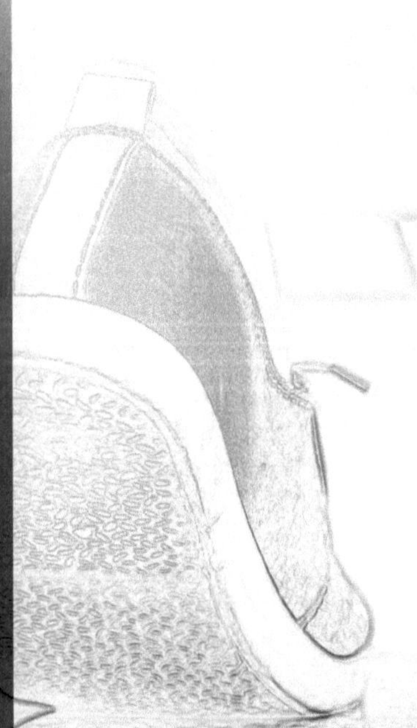

Occurrence Report



## Sustainable

### 5

All employees access and actively use information generated by occurrence reports.

The focus is on proactive reporting handling the behavioral deviation. Trend analyses indicate systemic and immediate actions.

High trust and commitment in the use of reports. Everyone takes care of each other.

## Proactive

### 4

People now report all occurrences (reactive and proactive). A trend analysis is conducted, and actions are taken. There is a search for the root cause. There are quality goals and follow-up of recommendations. All members of the organization are fit to report and understand the value of reporting.

The entire process is managed by the Safety Committee, and the data feeds the Company's Strategic Planning.

## Bureaucratic

### 3

An occurrence reporting system is established, considering all accidents, near misses, and unsafe conditions. The number of reports is what matters. Goals are set for the number of reports. The entire process is managed by the work safety department. Most reports are made by the leadership and safety team.
Employees have not yet adhered to using the tool, as they consider the process not very transparent with regard to discussions.

## Reactive

### 2

There is no formal system.

Occurrences of serious physical and property damage are reported and minimally investigated.

Employees are scared to report anything and think they will be punished.

## Pathological

### 1

Occurrences are not reported. People who bring occurrences are penalized and seen as incompetent.

To get to the sustainable level,
see the **ALWAYS** and **NEVER**
recommended.

---

## ALWAYS

- Report all occurrences;
- The process can be managed for safety, but the owners of the areas are responsible for enhancing its use;
- Perform qualitative analysis of reports;
- Maintain current communication of results;
- Focus on the quality of reports;
- Investigate all occurrences;
- Look for technological solutions to enhance the use of the tool;
- Maintain high credibility processes;
- In negotiations, seek the root cause;
- Reinforce the reporting culture;
- Enhance the focus on reporting proactive occurrences (behavioral deviations); and
- Train employees to understand the importance, why to report, what the reporting value is.

---

## NEVER

- Enter quantity targets only;
- Penalize people who report;
- Investigate only serious occurrences;
- Hide the information;
- Stop giving feedback;
- Stop turning reports into actions and practical knowledge for everyone in the organization; and
- Allow only the leadership and safety area to report occurrences.

# 18.  Occurrence Investigation

The investigation of occurrences is a reactive ritual and must have a trained team that leads the process.

Heroes need to be held accountable and their behavior investigated.

The main symbol of an investigation is the learning put into practice to avoid repeating events.

Below, I have separated some questions regarding this element for your reflection:

☞   Are all occurrences really investigated?

☞   How does the investigation happen?

☞   Who leads it? Who is involved in it?

☞   When do we understand that the investigation of the occurrence is finished? How is the learning process after an occurrence?

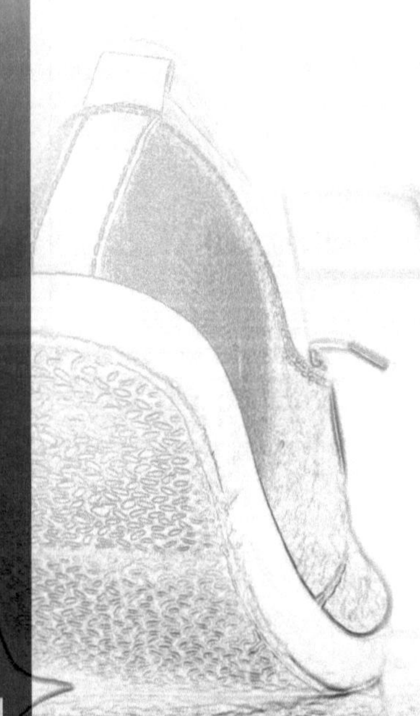

Check out the main aspects of investigating occurrences in each type of culture:

**Sustainable — 5**

Incremental improvements are made with the purpose of ensuring that the ritual is complete and that learning generates transformation and strengthening in the Safety Culture. The process is conducted by a multidisciplinary team crosswise and independently. The investigation takes into account the history of internal and external occurrences, and action plans are cross-cutting, adding innovation and optimization. Lessons learned are widely disclosed. Direct leadership related to the area of occurrence acts proactively reflecting on its responsibility and on what could have been done differently to stop the accident from happening.

**Proactive — 4**

Leadership is empowered and assumes its role in investigating occurrences. The root cause is found from a deep analysis of behavioral triggers and management failures. The immediate leader is now investigated and held accountable.

The safety area technically supports the investigation. Lessons learned are disclosed, but without discretion, and some areas may not receive the communication.

**Bureaucratic — 3**

All occurrences reported are investigated by the safety area. Lessons learned are developed and communicated only for accidents with high potential and impact on losses and damages. Corrective actions are focused on training and procedures. There is a tendency to define the root cause always as a behavioral deviation and to never involve and hold accountable the immediate leader.

**Reactive — 2**

High potential occurrences and those that have caused losses and damages are investigated by the safety area. There is fear when reporting accidents and incidents, as the focus of the investigation is to find guilty parties. There is no process for managing lessons learned from investigations.

**Pathological — 1**

Occurrences are not investigated or reported.

To get to the sustainable level, see the **ALWAYS** and **NEVER** recommended.

## ALWAYS

- With the prior information, communicate all those involved, according to the flow and standard of communication;
- Practice active listening to everyone involved;
- Conduct the investigation with a multidisciplinary group;
- Reconstruct the facts;
- Interview and build the empathy map;
- Conduct the process impartially;
- Look for behavioral triggers after finding unsafe behavior;
- Check the effectiveness of all implemented measures;
- Build on the lesson learned and disseminate it to the entire organization;
- Investigate all occurrences, regardless of their classification level; and
- Ensure that the people involved in the investigation process are trained.

## NEVER

- Look for culprits;
- Embarrass those involved;
- Be reactive;
- Analyze the facts unilaterally;
- Believe that only occurrences that have caused severe damage are worth investigating;
- Let safety professionals conduct or present an investigation (This is the responsibility of the Area Owner);
- Fail to report the occurrence, lesson learned and coverage actions;
- Allow the immediate leader to conduct an investigation of their employee;
- Jump to conclusions;
- Stop listening to those impacted;
- Conclude without risk analysis of the task;
- Fail to include human factors;
- Fail to investigate the immediate leader's responsibility in the occurrence;
- Stop consulting histories of internal and external occurrences that are similar; and
- Disclose lessons learned in a timely and isolated manner.

# 19. Emergencies

Emergency preparedness is a layer of defense within hierarchies of control. This preparation brings together practices found in symbols and rituals, such as: Emergency drills and all the work routine done with the firefighters.

The organization's heroes are supportive and always challenge existing emergency scenarios.

Here are some questions to think about:

☞ Are employees aware of possible emergency scenarios and how to act during possible occurrences?

☞ Does the business decision making have clear criteria related to investments to implement and maintain contingency measures and actions?

## Sustainable

**5**

Incremental innovations are made. Within the unit, 80% of employees and the entire unit leadership have been trained and were members of the emergency brigade. The site's entire middle leadership is part of the brigade.

The surrounding community is influenced by emergency management and preparedness. The emergency committee is a very active forum, and its actions transcend the factory gates.

## Proactive

**4**

The emergency preparedness and response program is widely disclosed, investments are prioritized and implemented, according to the risk matrix. There is an emergency committee who meets with a set periodicity to rediscuss scenarios. There is a structured communication process. The middle leadership comprises 25% of the unit's emergency brigade.

## Bureaucratic

**3**

The safety team together with maintenance and engineering and the site's leadership develops an emergency preparedness and response program, scenarios are discussed, measures are implemented. However, brigade members and the operational front line are not involved. The process does not have a structure for communication. Middle leadership is not part of the emergency brigade.

Competencies and meetings with brigade members are not in line with the emergency preparedness and response program and are treated in parallel.

## Reactive

**2**

There is a team of trained brigade members, and compliance with the local legal requirements related to preventing and fighting emergencies is implemented, but the focus is on emergencies resulting in fires and explosions.

## Pathological

**1**

There is no emergency preparedness and response system developed and implemented within the organization.

The company has a third-party provider who can act to contain a fire.

To get to the sustainable level, see the **ALWAYS** and **NEVER** recommended.

---

## ALWAYS

- Have an emergency preparedness and response program;
- Have the participation of senior and medium leadership;
- Senior and medium leadership must have training as a brigade;
- The emergency preparedness and response program must be built by a multidisciplinary team;
- Have an emergency committee with regular meetings;
- Update and test emergency scenarios;
- Have a simple-to-understand decision matrix;
- Build emergency competence across the organization;
- Invest in mitigating actions; and
- Communicate massively.

---

## NEVER

- Develop an emergency preparedness and response program based on reactive occurrences;
- Neglect training;
- Rationalize scenarios of possible emergencies, due to not knowing their occurrence;
- Let the program be the responsibility of the safety area;
- Develop the plan unilaterally; and
- Stop involving all areas and act in a multidisciplinary way.

# 20. Safety Communication

Safety communication practices must be transversal and permeate all levels of the organization, including all rituals and symbols.

It is a great supporter of heroes in the mission to inspire, engage and re-signify safety as a value within the organization.

Below, I have separated some questions regarding this element for your reflection:

☞ How, to whom and how often is safety communication done?

☞ What do we communicate?

☞ Who is the Spokesperson?

☞ What communication channels do we use the most?

☞ What kind of message do we deliver to our employees about safety?

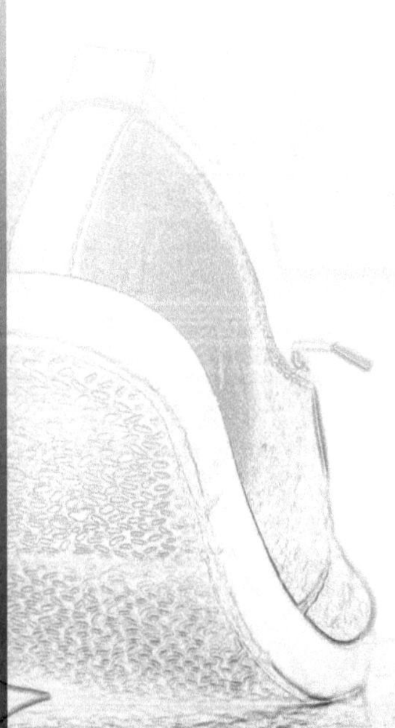

**Sustainable**

Communication is now integrated, where safety is the core message of the business. There is a structured calendar for the communication program.

The communication plan goes beyond the boundaries of the organization. Employees actively participate in the channels and rate shared contents as useful

**5**

**Proactive**

Spokespeople work in a bidirectional communication process on matters related to the safety system in the workplace. There is a structured calendar for the communication program. This plan gains prominence and presence in all media and reaches the entire organization with preventive and also thoughtful messages regarding internal occurrences for the purpose of learning.

**4**

**Bureaucratic**

A communication plan is developed together with the communication area, analyzing all channels and audiences. A minimum communication plan is prepared with a focus on the operational/production area.

The organization sets out who the safety spokespeople are. The process is a one-way street with little or almost no intentional listening.

**3**

**Reactive**

Topics selected by senior leadership are communicated reactively, after occurrences.

**2**

**Pathological**

No safety communication is carried out.

**1**

Check out the main aspects of safety communication in each type of culture:

To get to the sustainable level,
see the **ALWAYS** and **NEVER**
recommended.

## ALWAYS

- The safety message needs to be interesting, very clear and simple;
- Build a diagnosis of the communication channels used by the company (target audience, channel, language, and frequency);
- Establish a communication plan with frequency/content;
- Establish a spokesperson who is credible when talking about safety, who is an inspirational leader;
- Ensure that communication is two-way;
- Seek synergy with the business strategy and channels;
- Invest in simple, clear, sensitive, and appropriate communication to the public;
- Include in the communication all elements of humanized culture;
- Humanize communication, allow people to meet in the stories communicated;
- Put safety at the center of people's lives;
- Cause empathy and inspire people;
- Awaken people's intrinsic motivation;
- Spokespeople must have a genuine message about safety;
- The safety message needs to be interesting, very clear and simple; and
- Establish adherence KPIs in the planning of campaigns that speak to the learning matrix.

## NEVER

- Communicate reactively;
- Allow a silent safety program;
- Talk about safety in isolation;
- Work communication in waves and let it be perceived as "trendy";
- Stop establishing spokespeople;
- Communicate to embarrass;
- Make a bureaucratic or mandatory communication; and
- Stop paying attention to language and form, communication needs to be understood.

## 21. Driving with excellence

The uncontrollable elements related to accidents are in traffic, it is necessary to develop safe habits in people who frequent traffic within the organization.

Organizations need to internalize a specific care agenda, related to this topic, that in a transversal way, speaks to the culture maturation model.

**Sustainable**

**5**

The Safe Driving program is offered to all employees and their families. It is considered a safety benefit. It has a social focus. The organization has 3 consecutive years of significant reductions in vehicular safety performance, including Zero Fatalities.

The organization invests in technologies and researches that contribute to improving driver behavior and risk perception.

**Proactive**

**4**

There is a program serves those who drive for up to 20% of the work day. Management is done by safety. The entire leadership is engaged. Reactive and proactive KPIs have been established. Communication among participants is intense. There is an educational focus. At this level, tools are used such as Coaching Behind the Wheel, route risk analysis, and critical drivers from the evidence of psychosocial risks. The leadership of areas with drivers are ambassadors for the tools and fit to use them.

**Bureaucratic**

**3**

There is a vehicular program with a focus on training that only serves those who drive for equal to or more than 60% of the work day. The program is managed by the safety area. Little communication is carried out. Reactive KPIs are established. There is a punitive focus. In addition, the perception of messages received from the vehicular safety program is that the cargo and the company's brand displayed on the vehicle are more important than the driver's life.

**Reactive**

**2**

Serious vehicular occurrences are handled and a few actions are implemented reactively, always conducted by the safety area. There is no structured system for reporting traffic occurrences. No communication channels explore the focus on vehicle safety.

**Pathological**

**1**

There is no structured program for risks arising from traffic.

To get to the sustainable level,
see the **ALWAYS** and **NEVER**
recommended.

---

## ALWAYS

- Establish a Safe Driving program based on risk analysis of routes, drivers, and activities;
- Offer a safe driving program for all employees;
- Establish reactive and proactive KPIs;
- Allow yourself to multiply the lessons learned;
- Ensure clear and simple communication;
- Establish a sponsor as a leader of the organization;
- Introduce the benefits and value of a safety program having a strong social imprint;
- Establish a recognition program seeking to inspire and engage;
- Innovate and seek technologies that proactively support the evolution of the program, for example, the use of telemetry;
- Include the family and influence the community with safe traffic habits;
- Ensure that leadership is trained to be an ambassador for safe driving tools and habits; and
- Ensure that each leader applies Coaching Behind the Wheel to their employees at least once a year.

---

## NEVER

- Offer a program that has a punitive perception;
- Fail to involve and engage leadership;
- Fail to identify critical drivers through psychosocial assessments;
- Allow the program to be seen as yet another safety initiative;
- Believe that learning can only come from lessons and occurrences, work on proactive learning;
- Fail to educate, encourage, and inspire participants;
- Stop assessing learning adherence through coaching behind the wheel;
- Stop managing route changes based on the critical risks highlighted in the analyses; and
- Stop listening to drivers.

## 22. Performance Review and Continuous Improvement

Reviewing performance and seeking continuous improvement is a ritual of reflection, learning and taking action. Heroes must encourage this process in the pursuit of safety excellence.

To reflect, I have separated some questions below:

☞ Is critical analysis done when looking at safety indicators?

☞ What do we consider as input to the performance review?

☞ What is the performance review system? Who participates?

**Sustainable**

The process is conducted independently in the pursuit of improvement and excellence in Safety Culture. All involved understand their role and responsibility and act as "owners" seeking to build a continuous learning process. Feedback considers inputs from the community, sector, and even families of employees. The periodicity is biannual.

**5**

**Proactive**

The review process is managed by the safety team and led by the unit's senior and middle leadership. Proactive and reactive KPIs are established and converse with the unit's strategic planning. Results are widely communicated. Performance Review & Continuous Improvement are fed back by environment surveys and every 3 years by the Safety Culture diagnosis. The performance review periodicity is annual.

**4**

**Bureaucratic**

There is a structured performance review process, with an emphasis on continuous improvement and strengthening of the Safety Culture. The entire process is developed by the unit's leadership team through the safety committee. KPIs established are not tied to any other performance program of the organization. The process feedback is not structured. The process is communicated among the leaders and safety team. Results found do not evolve within the organization and are accommodated in action plans that move very slowly.

**3**

**Reactive**

There is no structured process for Performance Review & Continuous Improvement of the Safety Culture. A few actions are taken occasionally due to specific stakeholder demands.

**2**

**Pathological**

The safety performance review process is not carried out, and there is no focus on continuous improvement.

**1**

Check out the main aspects of performance review and continuous improvement in each type of culture:

To get to the sustainable level, see the **ALWAYS** and **NEVER** recommended.

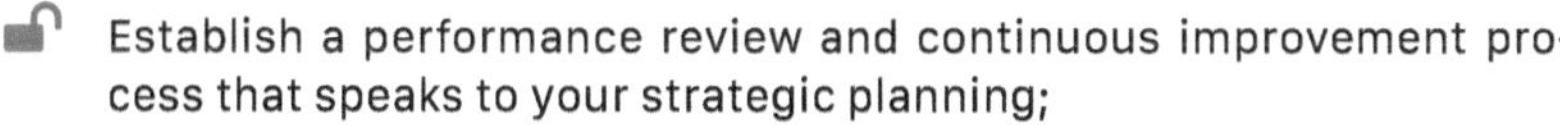

---

## ALWAYS

🔓 Establish a performance review and continuous improvement process that speaks to your strategic planning;

🔓 The performance review and continuous improvement process must be fed back by the climate survey and every 3 years by the culture diagnosis;

🔓 The review should be done by the multidisciplinary leadership team;

🔓 Proactive and reactive KPIs must be established;

🔓 The results of the process are widely communicated;

🔓 Conduct a process that results in continuous learning for the organization;

🔓 Communicate improvements widely across the organization; and

🔓 Continuous improvement needs to be evidenced through active listening from the workforce.

---

## NEVER

🔒 Allow the review process to be driven and developed by the safety team;

🔒 Stop measuring learning and mapped improvements;

🔒 Stop performing critical analysis of the performance review process;

🔒 Lose genuine passion for continuous improvement;

🔒 Enable Performance & Continuous Improvement review based on reactive KPIs; and

🔒 Disregard as mandatory inputs in the critical analysis the use of the following tools: committee, risk analysis, strategic planning, reactive and proactive indicators, occurrences, behavior observation, competence management, change management, process management, communication agenda and others that are linked to the safety culture.

## 23.  Process Safety Management

Process safety is a tool and permeates several rituals and symbols and needs to be on the Heroes' agenda.

We need to build reliable processes and constantly challenge maturity levels.

- Is there an efficient management of processes and are they really safe?

- How is the reliability of safety processes worked?

- How is process safety governance done?

- Is there a Process Safety Committee?

**Sustainable — 5**

The process safety committee answers to the organization's senior leadership, linking its strategic planning to the organization's strategic planning. Funds for process safety management are "exclusive." Extra funds are provided to solve unplanned operating issues, so this does not impact planning. There is a level of awareness of the importance of process safety management by all employees.

**Proactive — 4**

The purpose of process risk management is understood and multiplied by the entire organization. The process safety committee is established. Proactive and reactive KPIs are managed. The process risk management is linked to the organization's strategic planning. There is a risk basis in the maintenance plan. Trends are used to improve maintenance planning. If funds have been deviated to resolve operating issues, extra funds are obtained to ensure that the maintenance plan will be carried out. The action priority of the maintenance team is to carry out interventions with quality.

**Bureaucratic — 3**

A few process safety management guidelines are prepared, but most are not in the local language. Management is by the maintenance, engineering, and safety areas. The action roles, responsibilities, and pillars start being cascaded down as mandatory. Reactive KPIs are established. There is a system for monitoring maintenance, accrual, costs, and defects activities. The maintenance plan should be followed, but in reality its funds are deviated to solve operating issues. In addition, the focus is on keeping interventions to the least possible amount of time.

Priority is on producing and meeting goals.

**Reactive — 2**

Management is punctual after any occurrence with severe damages. The entire process is managed by the maintenance area. Critical equipment is maintained at the manufacturer's minimum guidelines. Otherwise, only urgent repairs are done. Past problems determine what is prioritize by maintenance.

**Pathological — 1**

There is no process safety management. This is not seen as something important. Equipment is used until it breaks down. Afterwards, it is temporarily fixed, and alternative services on defective equipment are common to keep the operation working. Priority is on minimizing costs and not halting the operation.

To get to the sustainable level,
see the **ALWAYS** and **NEVER**
recommended.

---

## ALWAYS

- Establish a committee to manage the pillars, roles & responsibilities, proactive and reactive KPIs;
- Have a plan with actions linked to short, medium and long-term process risk analysis;
- Keep the communication intense and sensitive to the topic;
- Involve the multidisciplinary team in the process safety committee;
- Engage the operational front line and sharpen employees' risk perception;
- Link process safety planning actions to the organization's strategic planning;
- Make the equipment manuals available in the local language;
- Ensure that the quality of maintenance interventions is prioritized over time; and
- Ensure that senior leadership is constantly made aware of the importance and status of process safety plans and actively participates in decision making.

---

## NEVER

- Rationalize process hazards and risks;
- Work in a reactive, punitive way and with reactive KPIs;
- Allow process safety concepts to permeate only maintenance, engineering, and safety staff;
- Restrict process safety tools to committee members only;
- Use maintenance budget for other demands;
- Stop prioritizing dealings related to critical equipment; and
- Ignore the manufacturers guidelines and manuals.

# 24. License to Operate

Just as a factory needs all governmental licenses to operate, when we think of our leaders and employees, and of the entire process that involves the maturation of the safety culture, the development of the competence matrix is fundamental, which must be understood as the license to operate for all our employees.

Building a safety competence is critical to sustaining the journey towards strengthening a safety culture.

This process must be recognized as indeed a journey and not a destination.

All employees must have their safety competence constantly developed and encouraged, and there must be no hierarchical distinction.

To support understanding, I prepared the questions below:

- What are the technical and non-technical competencies that exist today and which ones will we develop within the organization to strengthen safety?

- How is adherence to competency development verified?

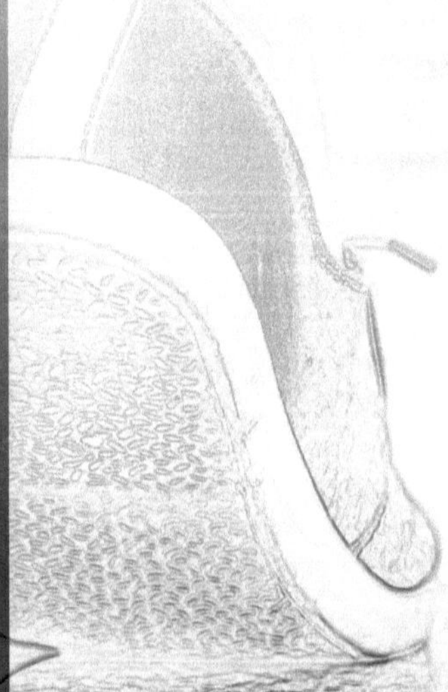

Check out the main aspects of license to operate in each type of culture:

**Sustainable**

**5**

The entire senior leadership has undergone formal training on Safety Culture. Additionally, the local leadership actively makes sure that the entire senior and middle leadership receives this and other relevant training programs according to the license to operate. Mechanisms are established to check the effectiveness of and adherence to lessons learned put into practice and this occasionally results in better behavior in safety leadership. It can be noted that the development of skills is seen as a never-ending process.

**Proactive**

**4**

All levels undergo proper safety training as part of their license to operate with regard to the technical and non-technical skills required for each function. And this includes all new staff and role changes. There is a formation of "Area owners/sponsors" multipliers to develop middle leadership in safety so that they have compatible competence and skills in cases of promotion. The license to operate converses with career path programs.

**Bureaucratic**

**3**

Competence matrices are present, and a lot of standard training is given. Knowledge acquired in courses is tested. Employees are interested in showing that they have participated in all required courses. Additionally, the unit identifies the needs for safety training for each manager/supervisor level and integrates into the license to operate as a prerequisite for performing the work.

The license to operate program does not integrate the career path, they are parallel initiatives. Training programs are seen as bureaucratic.

**Reactive**

**2**

After a serious occurrence, funds are made available for specific training programs, despite initiatives decreasing as time goes by. There is no structured training and development system that converses with career path.

**Pathological**

**1**

Safety training is seen as a necessary evil. They comply with training when required by law.

To get to the sustainable level, see the **ALWAYS** and **NEVER** recommended.

---

## ALWAYS

- Establish a matrix that proactively acts as a prerequisite for the development of any activity;
- The competence matrix must be systemic and speak to all the functions of the organization;
- Be a credible program;
- Develop skills with experiences that stimulate learning, based on the learning pyramid;
- Link the competency matrix to the career track;
- Update the skills portfolio by harmonizing technical and non-technical skills; and
- Keep the portfolio up to date in line with the core competencies outlined in the World Business Economic Forum publications.

---

## NEVER

- Develop a matrix focused on reactive occurrences;
- Be managed by safety to maintain the "entity" state of safety;
- Be bureaucratic;
- Improvise formations;
- Fail to comply with the competence matrix; and
- Allow an employee to start their activities without knowing their competence matrix and which are the technical and non-technical safety competencies necessary for the development of their activities.

FROM THEORY TO PRACTICE

# The Relationship between Culture, Leadership and Safety Performance

*Perfection is achieved, not when there is nothing more to add. But when there is nothing left to take away.*

Antoine de Saint-Exupéry

When we try to build and strengthen the safety culture, there are some internationally recognized references and theories. In my view, two of them are fundamental in understanding the practical concept of safety culture and how to keep this concept strengthened, energized and with a preventive and anticipatory bias within organizations. I'm talking about Frank Bird's Pyramid and the Behavioral Iceberg, which I'll explain below.

## Frank Bird's Pyramid

The American engineer Frank Bird Jr., in the late 1960s, disseminated a theory inspired by a study begun decades earlier, in the 1930s, by Herbert Heinrich. In 1931, Heinrich, also an engineer and an American, published the work Accident Prevention: A scientific study, based on the analysis of accident-related data collected by the company he worked for, Traveler Insurance Company.

For the next 30 years he continued to collect information and with it he was able to identify the factors that caused the accidents. Heinrich analyzed 75,000 cases to arrive at the parameter 1-29-300. That is, he concluded that for every serious injury, there were 29 minor injuries and 300 incidents.

Bird, who also worked for an insurance company, Insurance Company of North America, started from his colleague's publications and analyzed another 1.7 million accidents reported by 297 client companies from 21 different industrial areas. The result of the study was summarized in a pyramid that he named the safety pyramid (or accident triangle) and which is now known around the world as Bird's Pyramid. The engineer concluded that for every serious injury accident there were 9.8 minor injuries (requiring only first aid), 30 property damage events and 600 incidents (near misses).

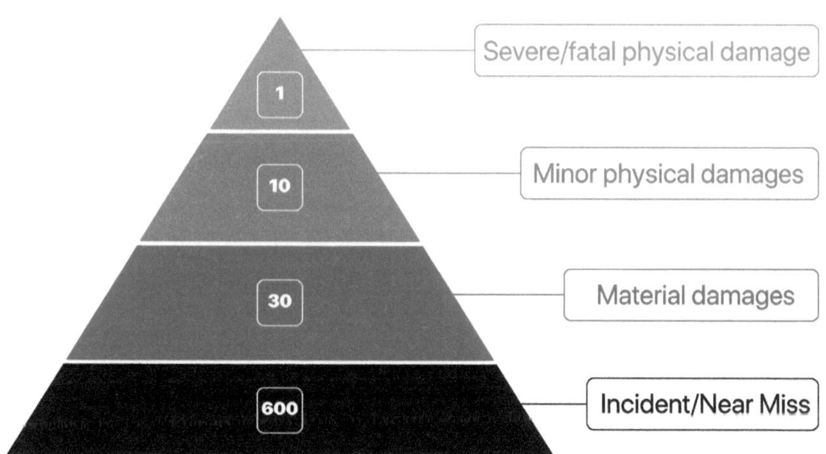


Figure 6:   Frank Bird's Pyramid

Both Heinrich's and Bird's pioneering work were considered innovative in their time, as they transversally managed to communicate and bring to the most diverse businesses around the world the statistical perspective of behavioral thinking and the prevention of work accidents. They traced an important relationship of frequency and severity between events (accidents and incidents) and, by showing that many accidents have common causes – both the most serious and the least serious –, they indicated that when addressing the most frequent, those without injuries, it would be possible to prevent the most serious ones.

Bird's theory, an evolution of Heinrich's, is still a reliable and inspiring source of reference for the safety agenda within corporations.

When we look at the peak of the pyramid versus its base, it's like saying that every major accident is a kind of book we'd rather not read. There were a number of previous warnings that were ignored, with enough information to prevent injury or even death, but which no one looked at and took action on.

Years after Bird's research, in the late 1990s, DuPont added a new statistic to the base of the existing pyramid, accounting for 30,000 misbehavior or unsafe acts committed for every fatal accident. Between the deviations and the fatal accident, there are still 3 thousand incidents, 300 accidents without leave and 30 accidents with leave. When we evaluate this proportion, it's impossible not to think that one of the reasons why no one read that book before something serious happened is, many times, our own ignorance.

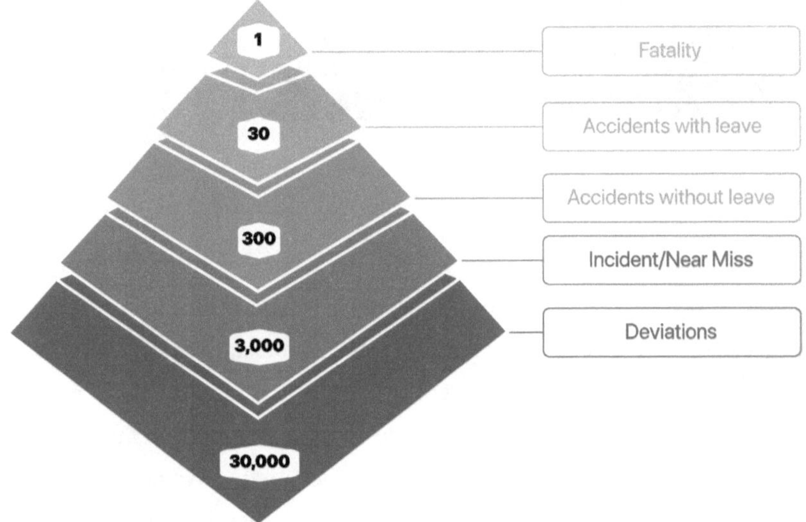


Figure 7:   DuPont Pyramid

The 30,000 behavioral deviations are almost always caused by: haste, improvisation, exception, presumption, and self-confidence.

**Haste:** speed is almost never synonymous with quality. Wanting to finish a task in a shorter time than it actually requires is a sure recipe for error – and the consequent risk of an accident.

**Improvisation:** it's that action linked to the well-known "Brazilian way" and the "let me take care of it." We must consider not only the improvisation of machines, equipment, and tools, it's necessary to pay attention to human improvisation when people are going to perform a task for which they were never trained.

**Exception:** how many times do we hear "just this once," or "just today," when performing certain actions? Whoever makes an exception becomes its slave.

**Presumption:** it's thinking you know, wanting to show you know without really knowing, it's the shame of assuming you don't have full knowledge about something, it's not wanting to ask for help. Presumption goes hand in hand with self-confidence.

**Self-confidence:** is the "I'm sure this won't happen to me." Self-confidence is always influenced by what we already know, experience, and expect, and excess of it can lead to reckless attitudes.

It's important that we see the safety pyramids not only as a statistical model full of numbers, but as tools for a more practical reading and understanding of the safety culture's foundation.

When I evaluate the top of the pyramid, which refers to reactive indicators based on the occurrence of accidents, for example, I understand that it points out the relationship between the frequency and the severity between them. However, this frequency and severity of accidents reference, beware, is always a consequence of the level of maturity of the safety culture existing in that place.

In this deeper interpretation, using my experience in safety culture in different companies and projects, I built my own pyramid. In it, the bearer and/or owner of the culture is the immediate leader, the most important figure because they carry within themselves the transforming power to engage, encourage, educate, inspire, and form teams sharing their values and beliefs, so that they can anticipate and prevent the effects of triggers and behave safely. I explain below what each of the levels in the model represented is, and in this way it becomes possible to understand the relationship that is the subject of this chapter.

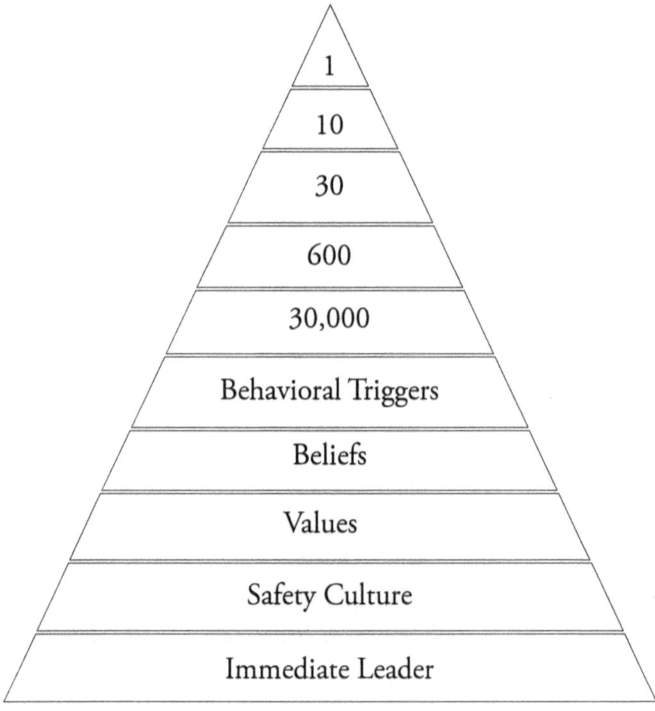


Figure 8: A model representing the relationship between culture, leadership, and safety performance.

## Behavioral Deviations

The journey to build this relationship between Bird's Pyramid and my model starts with dealing with behavioral deviations, also known as unsafe acts. We have already dealt with them a little further above. Before them there are the incidents and/or near misses, represented by Bird, events that do not cause physical damage or loss, but that, after they happen, show their potential severity, if their occurrence had materialized. These little scares that we go through on a daily basis are usually accompanied by noble justifications: it was haste, improvisation, exception, I assumed it was like that and so many others. That slip without a fall and without injury, the one that was nothing more than a scare, and which you soon moved on from.

Behavioral deviations, also known as unsafe acts, are those that take a different path than what was established for the execution of a task, they are shortcuts to procedures and operational guides, improvisations of people, machines, equipment, and tools. Behavioral deviations are also linked to presuming. "I think I know" is dangerous and often leads to unsafe acts – and for me, it's a different way of saying "I don't know" whenever a person wants to sound smarter or less uninformed. However, this use of self-confidence, which is nothing more than an attempt to optimize time and frequency in the execution of tasks (i.e., haste), can cost lives. Pay attention to the fact that the increase in self-confidence is almost always proportional to the longer time in the company or in a role, and the employee's level of experience.

## Behavioral Triggers

Behavioral deviation is the first representation and/or demonstration of the safety culture with a practical action. And it's always a personal choice.

When I talk about this in my training, I see a shock expression on people's faces. Yes, the choice is yours! But then, I explain that behind the deviations there are behavioral triggers, activators directed by our thoughts, feelings, and emotions and which, precisely because of this, tend to be unstable. And not completely rational. I consider these triggers a very important topic in building a safety culture, so I dedicate the entire next chapter to delving deeper into the topic and the four elements that these triggers fit into.

## Beliefs

Beliefs are the result of all the ideas you saw, heard, or concluded that ended up becoming an absolute truth for you. In everything they do — the way they think, feel and act — individuals are driven by their beliefs, which is precisely why many people act differently in identical situations.

Beliefs are like magnets: you believe in a truth and it becomes real. If you believe that doing it safely is harder and/or will take longer, then that becomes true.

Beliefs are cognitions about the world, subjective probabilities that an object has a particular attribute or that an action will lead to a particular outcome. They can be clearly and unequivocally false.

In my view, our greatest challenge in transforming and strengthening the safety culture is precisely in our beliefs. A belief cannot be destroyed, as it tends to be deep and part of our set of convictions. When we contextualize beliefs, when it comes to safety, we can see that many of them are limiting, as they do not allow people who have them to evolve and manage to find a new way to act safely.

One of these days I was at an airport, waiting to board, and I started to think about the potential of beliefs, and the large number of beliefs that limit the strengthening of the safety culture within an organization. I took a paper and a pen and started to list the beliefs I sometimes found in teams I've worked with. In a few minutes, I reached 51 (see the complete list in the appendix of this book). I'm sure they don't exhaust the subject, but they show us their importance and why we can't ignore them. We are facing an extremely arduous challenge because we know that re-signifying and/or re-editing beliefs is a complicated task, because they are personal. There are many "attachments" to them that we need to question daily, showing why they are not sustainable, especially when it comes to safety.

And how is it possible to re-signify beliefs? In order to assess how we are able to engage, encourage and involve people, I would like to talk here about the golden circle model, also called the circle of trust. The way to re-signify a belief, in my view, goes through it – it is the proposals of this golden circle that give us an initial direction on how we should treat beliefs, put them on the table and use our relationships, influence and rituals to give new meaning to what limits people's safety.

The circle of trust concept gained popularity years ago with author and speaker Simon Sinek. For him, this circle is one of the most important changes in mindset that every organization must go through. Sinek sums up his idea with one basic sentence: "If you hire people just because they know how to do their jobs, they will work for your money. But if you hire people who believe what you believe, they will work for you with blood, sweat and tears." In other words, motivation must be the result of inspiration, not manipulation. And that's not the kind of conviction we need when it comes to safety.

The golden circle sets out to explain why some leaders and organizations have achieved such an exceptional degree of influence. And it shows how some leaders are able to inspire action rather than manipulate people into action.

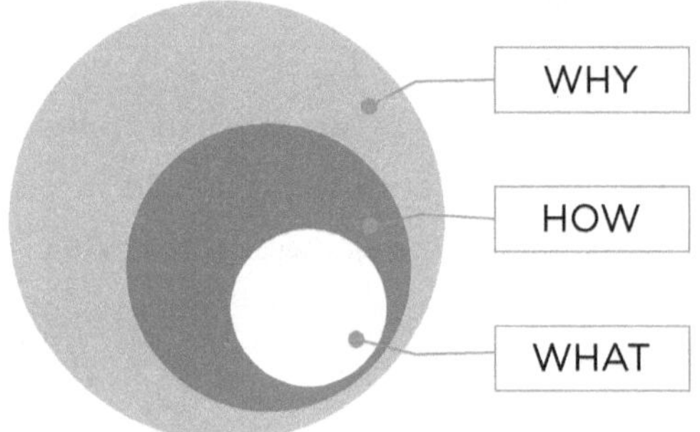


Figure 9:  Golden Circle Model.

Reading the circle starts from the inside out and works very well when it comes to safety and safety culture.

Ask yourself **why.**

Ask yourself **how.**

Ask yourself **what.**

It is from the use of this set of questions that we can re-signify limiting beliefs. We need to offer people the "why" Safety, and this why must be loaded with purpose, belief, mission, value, advantage, and conviction. After the why, we must build the "how" to do it safely from now on. And it's no longer safety first, it's about doing it "with" safety regardless of where and whoever you are, it comes down to understanding and applying that: all tasks can be performed safely. It's understanding that there is a reason for every procedure we follow. And finally, we deliver the "what" is nothing more than expectations and their individual role in the face of belief and its re-signification.

## Values

Values are ideals, guiding principles of life, or overarching goals that people strive to achieve. They direct human behavior and govern all of our decisions. And they are personal and non-negotiable.

Values are built throughout our lives, mainly through our parents or the people who raised us as children. In addition to education, at home and at school, there is the influence of our culture, the city where we live, the life experience, the family, the stories lived by our parents, grandparents, great-grandparents and so on. Values rarely change. On the contrary, they tend to be reinforced throughout our lives.

Our customs are based on values, they are emotional states to which we give importance and that we seek to experience, even if unconsciously.

They are also the values that connect us with our team and company. There are two main forms of connection, the first is intrinsic motivation - when the company's values speak directly to our personal values, for example when we treat safety as care, and this can be felt and perceived in the behavior of employees, in relationships, those people who have the same values will join it – and the other one due to extrinsic motivation – this one works very well for that audience that has a grateful relationship with the company where they work. I know people who are extremely grateful because it was in that company that they had the opportunity to learn, lead, earn money, buy their house, educate their children, etc. These people are unconditionally committed, even without fully understanding what it is about. The grateful represent 1% of our population. Other people who do not fit the characteristics detailed above can be extrinsically motivated with recognition and reward programs.

Values are responsible for engagement and inspiration within the organization. To achieve the values of each employee, it is initially necessary to understand what they are, what we are dealing with. To then understand what needs to change to re-signify safety. It's essential, in this exercise, to leave the traditional and the punitive. One should inspire, not compel, or punish. Only in this way can we captivate our employees and achieve adherence to our value called safety - which in practice needs to be understood as care.

## Safety Culture

Safety culture is still seen as something institutional, or rather, an entity that is represented by the organization's CNPJ (Corporate Taxpayer Registration). Our attempt with this book is to show that the safety culture is an integral part of personal/individual values and beliefs and this means that it needs to be understood as CPF (Individual Taxpayer Registration) referring to the citizen.

It is a demonstration of our choices and habits, based on the relationships of affection we have with our values and beliefs. And it is more stable than the safety climate, which means that we need time to work on strengthening it and increasing its maturity level.

When an accident happens in a unit, everyone is uncomfortable, wants to help, improve safety, the topic affects relationships, priorities, agendas, and deliveries of the moment. However, after about three months, the accident is forgotten and everyone goes back to doing things the same way they did before. That's climate!

The organization's culture is not affected by an accident. It can be a kick-off, but only with the support of the actions will we be able to verify the evolution of the maturity level on that particular risk.

## Immediate Leader

The one who brings the safety culture to the work environment, directly reflecting on habits and personal choices. The one who determines the tone. Their behavior in rituals and in the use of symbols is what strengthens or weakens the safety culture. The way they communicate safety reflects the meaning and importance it has.

Once we understand this, it is clear that it is not a cliché to have leaders who are leaders in safety, which means acting preventively with a focus on anticipation.

Safety leaders ask more questions than they answer, they question positively, stimulating thinking about failures. Any immediate leader can be a safety leader, practice, encourage and educate their team every day, every time, everywhere.

They question their own behavior when an accident happens to a member of their team. They ask, for example: What could I have done differently? Why didn't I see this before?

FROM THEORY TO PRACTICE

# Rituals, Heroes and Symbols that Shape and Materialize Culture

*Transforming starts with making the rice and beans of safety in an irreproachable way. Our plate contains the operational, the tactical and the strategic. And we need to eat the entire plate.*

Andreza Araújo

When we talk about transforming an organization's Safety Culture, I usually talk about three key points in my training. But, before delving into them more deeply, it is worth noting that transformation is always about what we keep doing, what is maintained, what lasts. It can, of course, be real and immediate – an action or resolution is enough for it to happen –, but its validation only takes place through constancy.

Raise your hand who has already witnessed changes in policies, norms, and conduct, but saw everything return to what it was before, after a few months (or even weeks)? Changing behaviors, including our own, takes self-discipline. And it, in turn, like the strengthening of a muscle, requires repetitive movements. Human beings are made of habits.

That's why, for me, it's impossible to talk about cultural transformation without talking about rituals. They are the fundamental starting point for everything to happen. Contrary to what common sense imagines, these rituals don't need to be time-consuming or expensive, they just need to be consistent. It could be, for example, starting each meeting with a few meaningful words about safety.

The three key points I mentioned above are directly linked to the construction of these rituals. They are:

**COHERENCE:** What I say must agree with what I believe in! We need people who make things happen, who take a genuine stand, reflecting their beliefs, purposes, and convictions. There should be no distance between what I say, what I think and what I believe.

**TRANSFORMATION VALUE:** It is very common in organizations for me to see that leaders put exposure above real cultural transformation when it comes to safety. In the search for the immediate result, without realizing it (or sometimes even consciously) we create monsters that do not bring transformation. A classic example are the top of pyramid indicators. They worry me a lot. Saying that "we have been accident-free for 1357 days" cannot be the main measure of strengthening a safety culture. I often see exaggerated pressure on the medical department, for example, to hold this number, often misclassifying events, so that this indicator remains untouched.

When we choose this path, we end up staying on the surface, without experiencing or bringing about transformation.

**COMMITMENT:** Being committed is key. But not with excuses and limiting beliefs, with the fate of the organization. There is a clear direction, we know where we want to go and the journey is arduous. Many of us started, others continue on the trail and the cycle lives on. It is essential to ensure that we understand our individual role in the transformation and the details of the route we are going to follow.

Our excuses, more often than not, reflect our limiting beliefs or frustrations. We must re-signify, move forward, aim at examples of determination and overcoming. Sports are full of them, like swimmer Michael Phelps. His first Olympic competition was in Sydney 2000, at age 15. Phelps returned home without medals. Four years later, in Athens, he won six gold and two bronze medals, and today he has 28, in the Olympics alone, and more than 30 world records broken. What Phelps would likely tell anyone is that losing was never an excuse for not trying again. On the contrary, he invested in different characteristics to avoid them – and I list them below.

**Learning from his own mistakes:** before competitions, he watches videos of his previous races, to identify flaws and analyze details on how to improve.

**Training:** no one is able to achieve excellence without investing time and effort. There will always be some pain and sweat when working hard, but in the end it will all be worth it when you reap the results.

**Set clear goals:** every day when he wakes up, Phelps outlines his goals and determines what he needs to do to achieve them.

**Prepare yourself mentally:** yes, the mental part is very important in this process. One's head needs to be prepared to follow plans and redraw routes if something unexpected comes their way.

**Persevere:** Phelps often says that his focus is not exclusively on the goal, on the result itself, but mainly on the process, on how to get where you want to go. And that it is essential to knowing how to take advantage of everything that leads you to your biggest goal.

This means that our excuses, even the most noble and true ones, need to be turned into good lessons, lessons that can (and should) be shared throughout our careers. If I WANT transformation, I need to start by MAKING one. And this transformation is born within me and will always be recognized as genuine.

## Teaching an adult to ride a bike

In my training, I use this metaphor a lot to translate the safety culture learning process. We know that learning is not a one-off event, but an ongoing process.

According to andragogy, which studies the learning process of adults, there are practical advice for making this path more efficient. And some of them may be safety related:

- "What is the value of me learning this?" Adults need to know the need to learn something before they start learning.

- Adults have a self-concept of being responsible for their own decisions, for their own lives. This means that they resent and resist situations in which they feel others are imposing their will. We need to create learning experiences where adults move from dependent to self-directed learning.

- Adults thrive in environments that have a high volume of quality experiences. In practice, we must provide activities that explore the sharing of experiences as elements to provide a reflection on the habits and prejudices that learners have.

- Adults learn the things they need to know, the things they need to be able to do in a real situation, that is, the things they will use in subsequent days.

- Adults have their learning centered on life and their daily dilemmas. In this sense, we need to encourage learning using everyday examples, applied to the routine and real life of our employees.

- Adults respond to internal and external motivating factors ranging from recognition practices to feeling cared for in their work environment.

## The importance of rituals

There are many rituals within organizations that help to promote transformations. Once again, I see them working as a kind of solar system, where each planet has its importance and characteristics, without a defined priority. Each of the rituals, in the same way, contributes in a unique way to the construction of a sustainable and interdependent safety culture. We will see below what they are.

## Recognition & Reward

Within organizations, recognition & reward programs are rituals that strengthen the safety culture. But is all recognition and reward really being applied properly, at the right time, by the right people?

Why is recognizing and rewarding important? The concept is linked to the theory of expectations, a cognitive theory that relates the effort made in different tasks to their execution or performance. Recognition, I have no doubt, is a key element of each employee's relationship with their work and with the organization, especially when it comes to safe choices. And it needs to be integrated into the organization's culture.

What I see, however, is that people tend to expect grandiose programs with diverse and complex selection criteria when it comes to recognition. I remember a meeting I once attended to point out, within a company, the winner of a continuous improvement award. I left that meeting embarrassed, because the grades I had given to the participants were very high, but none of them managed, from the other members of the committee, to score higher than 5. It's as if nowadays, for someone to be recognized, they need to create a new iPhone every day.

We live in a moment in which recognition has become something extremely inflated. When, in my view, due to the power it has to leverage and transform organizations, it should be more commonplace, be in genuine "congratulations," accompanied by phrases such as "how nice to have you on my team," "I am proud to work with you," "today you inspired me," or simply "good job" or "great idea."

The power of CONGRATULATIONS is undeniable, simply because we cannot buy it at the bakery or supermarket. But when we get one, it changes the direction of our day. It is up to a good leader to recognize the work of his team, who are there, giving their best, without reservations. We cannot assume that they are just doing what they are paid to do.

Many leaders tend to believe that implementing a safety awareness program requires a specific, high budget. But the great news is, no, it doesn't have to cost a lot. And if it's part of an effective overall employee safety program, it can be cost-effective and generate a positive return on investment by helping to reduce the number of workplace accidents and the resulting costs.

Below are important components of an effective employee safety recognition and reward program:

1.    Permeate all levels of the organization;

2.    Have short deadlines, and with short duration;

3.    Promote moments of enjoyment with family and friends (the rewards);

4.    Be part of the company's cultural transformation program/agenda;

5.    Have a spokesperson who is an important company executive;

6.    Have SMART (S-Specific, M-Measurable, A-Achievable, R-Relevant, T-Temporal) performance goals;

7.    Celebrate success.

"Great companies make extraordinary use of celebrating the winner when it happens" is a valuable quote from the famous book The Pursuit of Excellence. We are not Super Bowl players, nor are we looking for the three-pointer that will win an NBA championship. But work is a big part of our lives, and in many ways, it's our own Super Bowl. Therefore, we must treat it as such and reinforce and recognize our employees when they achieve and maintain the required levels of safety.

## There are some practical tips that organizations can put into action to make the program more efficient:

- Invest in continuous, fun, visible and colorful communication about the program. Choose a theme. Work with posters, banners, newsletters, website, voicemail, CEO corporate email messages, messages printed on paycheck receipts, etc. Hold a fun launch meeting, with music, balloons, pizza for lunch. Hand out freebies such as company logo pens, beverage bottles, t-shirts, and advertise the program by explaining it in detail.

- Build reward components such as gift cards, merchandise awards, etc., with the general themes of the safety program printed on them, to reinforce the message. Gift cards are often popular because they allow employees to choose their own reward. They are cost-effective for employers, easy to distribute and present.

- Create training classes, promote fun, and recognize and thank employees for attending, for participating. Gamification is a tool that helps a lot in this way (read more about it below). Announce, during these classes, that quizzes will be held and that the first to answer the questions correctly will receive a gift card. Another possibility is to grant a special gift card, a kind of "mystery prize." Consider dividing classes into teams for fun competition. The more fun your safety training class is, the more your employees will learn about safety.

☞ Conduct quarterly safety tests or quizzes with those who achieve a specific score and hand out prizes.

☞ Catch someone doing the right job. All supervisors, plant managers and general managers should look for opportunities to catch employees who are working safely, wearing the right equipment and safety clothing, and following safe work practices. Recognize the employee on the spot, in front of their peers.

☞ You can establish a more formal employee safety recognition program – individual, in teams, shifts or for the plant as a whole.

## Gamification and its potential in recognition and reward rituals

Currently, games are everywhere, at all times we are invited to become loyal, to collect points, to exchange virtual currencies. In this scenario, how to take advantage of the power of games to massify the safety culture within organizations?

First, let's try to understand what gamification means. It's the idea of using game elements and mechanics in order to increase participation and generate engagement in everyday activities. Games promote immersion, attention focus, sustained dedication, creativity, strategic thinking, and engagement. In addition, according to some authors, they make us happy and provide a sense of well-being.

Games are being used in different areas – from education to the army. The obstacles inserted challenge our abilities and demand creativity, making the activity extremely interesting and efficient.

## Other advantages of gamification are:

☞ Games have well-defined goals, making the purpose of the activity clear.

☞ Games have rules, obstacles placed in the way.

☞ Games provide feedback with points, levels, leaderboards or even a progress track.

☞ Players have voluntary participation, but each participant must accept the objective, rules, and feedback system.

And how to use them specifically in recognition?

Promote competition: award winners for the best performance on a given date.

Recognize with medals.

Reward with promotions, loyalty program style. Example: gather ten coupons and exchange them for something.

When talking about gamification and its advantages, I think of the case of The Speed Camera Lottery, from Sweden. The main idea is to encourage drivers to obey speed limits in a fun way. The concept was implemented in November 2010, in Stockholm, by the Swedish national society, traffic safety and The Fun Theory (a Volkswagen initiative). As with any electronic radar, violators are caught by cameras and fines are generated. The difference is that drivers who obey the speed limit are also photographed and identified, competing for a lottery financed by a cash fund formed by the fines of violators.

The mechanism used is reward and managed to change the drivers' behavior: during the three days of the experiment, 24,857 cars passed the radar and it was found that the average speed in Stockholm dropped from 32km/h to 25km/h, that is, a 22% reduction.

Proof that fun is a powerful tool to change people's behavior for the better!

# 1.   Safety committee

The committee is the main forum for discussing, developing, and approving topics on the unit's safety agenda. They bring together leading representatives from all areas in a common goal: Strengthen the safety culture. For this reason, it's important to keep up to date with our meetings, which are also considered a safety ritual by leaders.

The premises of an effective safety committee are: to guarantee transparency in the discussion and dealings, visible commitment of the members, balance in dealing with emergencies and importance and, finally, a strong sense of ownership by its members, manifested from the constant search and genuine way of energizing safety with new experiences and stimuli.

The committee, in addition to having in its meetings the manifestation of a ritual, is also a symbol, a heritage of its unit and, like its members, it must serve it in its journey towards zero accidents and culture strengthening.

# 2.   Gemba Walk or Safety Tour

It's a visit, with a periodic schedule, which must be carried out by the senior and middle leadership of the unit or the organization as a whole, with the objective of understanding how safety is being practiced, with a view to new and possible opportunities for correction of deviations. The tour can also serve for behavioral observation, in addition to being a moment to promote informal recognition.

# 3.   Behavioral observation

We will talk more in depth about the topic later on, but in general terms, it means to invest time in, as the name itself explains it, observing the behavior of employees to, from it, generate conversations aimed at building a safety culture, providing guidance on deviations, and strengthening good habits. It's a proactive safety tool.

## 4. Periodic medical examinations

Occupational health ritual that aims to ensure the health of each employee. It follows a legal periodicity that must be respected and shows the care that the organization has for its employees. In addition to directing occupational health programs.

## 5. Integration

It's every employee's first safety ritual. In many cases, it's the only opportunity to talk about the topic with the employee and, therefore, the moment should be taken so that they receive all the safety guidelines about their tasks and the operation as a whole. In addition to being a moment to reinforce risk perception. It's important to emphasize that integration is a dialogue and not a monologue, and its adherence needs to be verified.

## 6. DDS or DSS

Daily Safety Dialogue or Weekly Safety Dialogue was the first established safety ritual around the world. They should be quick – the suggestion is that they are daily and do not take more than 5 minutes of work time. These dialogues bring a reflection related to safe behavior and are a stimulus to the perception of risks.

I strongly believe in the importance of this ritual and I often compare it to this metaphor:

*When we are on a road and come across a cross, usually located in a dangerous curve, the first reaction is to stick both hands to the steering wheel, followed by a slight decrease in speed. We feel the danger, the pain, and the fear, so we decide to be more attentive.*

*DDS, as well as the cross, needs to generate a reflection, which, even though fast and unexpected, can produce change in us, bringing awareness to each worker.*

*The DDS driver needs to understand that the words we speak have power. They can hurt or help, encourage or discourage, set someone up or bring someone down. Our message as Safety Leaders needs to be loaded with positivity, proactiveness, hope and possibilities. So adjust your focus and elevate your message. Let's positively impact people's lives. May our speech be a constant planting of life!*

*The characteristic of an effective DDS is that it needs to be strong, fast, point blank. It cannot have a bureaucratic tone, the one focused on the attendance list. So that, when the worker returns to their job, they are inspired, present, focused on the task, perceiving details, and anticipating risks.*

*We need to be bothered with Dialogues where one person is reading a piece of paper, being monotonous, the whole team distracted and preoccupied with signing the attendance list.*

*Attendance lists do not prevent accidents, I understand that they're important, but being content with paper puts us closer to an accident/incident*

## 7.  Emergency brigade

Group formed, in most cases, by volunteer employees who are willing to learn about preventing and combating emergencies. There are regular meetings and training, in addition to an inspection routine.

## 8.  Work permission

Lifesaving ritual used to validate and approve high-risk work. A risk analysis based on a specific checklist for the types of hazards involved in the activity evaluated, with a questionnaire detailing and controlling the team's exposure to hazards and risks.

## 9.  CIPA

Internal Commission for Accident Prevention, it's a legal ritual and, therefore, mandatory. It brings together a group of employees elected by the employees themselves and appointed by the unit's senior leadership. This group's mission is to work on accident prevention, with periodic meetings and functions recommended by law.

## 10. Mandatory training

They are also established by law and are mandatory. They aim to stimulate risk perception and build technical safety competence in employees.

## 11. Audit

Continuous improvement ritual that allows a self-analysis of everything that has been done and experienced safely in the unit. At the time of the audit, a comparison analysis is made between the theoretical and market standards established and the current practices implemented.

## 12. Benchmarking

Meeting with the sector and/or different sectors of the economy that also offers an opportunity to learn about what is being done in the rest of the market and in the world in relation to safety.

## 13. Trainings

They are established based on strategic safety planning, at intervals defined by the unit, with a focus on encouraging risk perception, reinforcing safe behavior, good practices and building technical competence in safety.

## 14. Investigation of occurrences

Ritual that involves several areas with the objective of seeking the root cause of occurrences, to establish an action plan so that they do not recur. It's important, of course, to build on a lesson learned from the occurrence and cascade it throughout the organization. The ritual will only be complete when comprehensive actions are implemented and prove to be efficient, in order to ensure that there is no repetition of occurrences.

## 15. Safety moment

Similar to DDS but happens before the start of various organization meetings. It should also be a brief reflection, without taking too much time.

## 16. Feedback

People management ritual that should also absorb comments related to safe behavior, including safety habits and attitudes. It's important that, based on the feedback, an action plan is actually built for the construction and continuous improvement of the safety culture, reflecting the safe behavior of the employee. Feedback is one of the rituals that allows safety to become part of an employee's career path.

## 17. Inspections and checklist

They can be of the most varied themes and are tools to stimulate risk perception, to identify unsafe conditions and opportunities for improvement. They must be carried out by an experienced and properly trained team.

## 18. Risk analysis

It's the heart of the safety management system and serves to quantify and classify existing risks. It can be done by location or by activity, and the main variables that help in the classification are: frequency of exposure to the hazard, severity of occurrence and probability of occurrence.

## 19. Work order

Legal ritual and serves to guide the employee, according to the tasks they will have to perform, about the risks to which they will be exposed and the control measures to avoid them, the PPE's, and the expected safe behavior.

## 20. Safety blitz

Normally developed by CIPA, it serves to reach a large number of employees by developing in them the safety behavior that you want to achieve. It's a catalytic ritual and a lightning event, performed without warning to maintain the surprise factor.

## 21. Sipat

International Week for the Prevention of Accidents at Work is also a legal, mandatory ritual, with the aim of preventing accidents. It's one of the best-known rituals in Brazil and one of great internal mobilization in companies. Most of the employees participate, being able, in addition to lectures, to mobilize with playful actions, stimulating safe behavior. The Brazilian legislation recommends some topics that should be part of the agenda.

## 22. Annual goal plan

It's an integral part of a continuous improvement process, where the main safety indicators are evaluated, both internal and external performance, with the objective of establishing goals that are challenging and achievable.

## 23. Critical safety review

This ritual is where safety performance is evaluated, through the stratification of the most diverse KPIs at the top and bottom of the pyramid, with a more detailed scope and, therefore, with deeper results. Conducting the safety review is an integral part of the annual target plan.

## 24. Campaigns

An annual calendar should be built with the topics that will be worked on in order to raise awareness and bring about changes in behavior. They will serve as a stimulus to risk perception and care. It's the execution of communication practices.

## 25. Safety day

April 28, a date established by the ILO, is a day to promote the prevention of occupational accidents and diseases around the world. It's an awareness campaign designed to focus international attention on emerging trends in the field of occupational safety and health and the magnitude of work-related injuries, illnesses, and fatalities worldwide.

## 26. Occurrences report

Reports of accidents, near misses and unsafe conditions must be carried out with the aim of giving visibility to the root cause and ensuring that each of them has mitigating actions so that they do not recur.

## 27. Change management

It's a discipline that guides how we can prepare for changes, both physical and in people, and should bring an approach to support the organization in these processes. Changes must be made in a way that alters and contributes to strengthening the climate and culture.

## 28. Shift change

Operational ritual that needs to include safety, a quick moment in which there must be, among the teams, an exchange of general information about the performance and safety events of the operation.

## 29. Lessons learned

Ritual arising from the investigation of occurrences, these lessons must be communicated to all areas, and not restricted to only those where there has been or there may be a similar occurrence.

## 30. Senior leadership visit

The organization's senior leadership (directors and presidency) must promote, at intervals to be defined, a visit with the sole purpose of improving the safety of its employees. This visit is a visible demonstration of commitment.

## 31. PPE delivery

Personal protective equipment is mandatory by law, but its handover ritual must be taken advantage of. The way it is done and the guidelines that are delivered with it make the moment a strengthening of the safety culture.

## 32. Emergency drill

It consists of training that simulates, in a real way, a risk situation in the company, making all employees follow the same steps. Emergency exits and available safety equipment must be used, checking their utilities and operation. For employees, it is also essential to understand the basic steps and concepts on the topic.

## 33. Health, safety, and environment management system

A set of procedures for managing or administering an organization, in order to improve the efficiency of the management of occupational safety and health (OSH) risks and impacts on the environment, related to all activities of the organization. Currently, management systems are based on international standards, such as: ISO 14000, OHSAS 18000, and ISO 45000 series of standards. The involvement and participation of each employee in the system implementation process is of fundamental importance.

## The importance of heroes

In the safety culture, heroes are spokespersons, multipliers, and safety bearers. In practice, they are the immediate leaders, and they translate safe habits, beliefs, and values through their attitudes. We will deal with leadership in a specific chapter of this book. But I want to reinforce its importance here with a real story.

I once had the challenge of drastically changing safety numbers within a large company. The goal was aggressive: to reduce, in 180 days, the number of occurrences by 50%. For that, I was sure, it would take genuine engagement from the senior leadership. And that's what happened. The vice president of operations embraced the project with tooth and nail. He was a living example of safe behavior and spoke about the topic with passion and truth. We achieved the proposed goal, and so I wanted to know from the executive what had made him such an effective safety spokesperson. And the answer I heard was. "A sad personal story."

More than three decades before our project, a car accident had claimed the life of his only baby boy. He was returning from a party at the factory – at a company he worked for –, his wife and son in the front seat and without a seat belt (at that time the use of car seats and seat belts were not mandatory), and their car had been hit head on. The child died instantly. He and his wife, who spent more than ten days in a coma in the hospital, could not even bury their precious son.

It's unbelievable the transformative power that a story can have in a person's life – and how much the change in behavior it entailed was responsible for the formation of the great leader in safety that that executive had become. After the accident, he had written a letter to other driving parents, and whenever he saw any child in the front seat of a car, he would stop and deliver that letter. "My mission became to save other lives. My tragedy brought with it a value that not even a million rules could have built."

What we get from this episode, the courage to speak openly about it and use one' own story without fear of exposing oneself to save other lives is the necessary break to live safety as a value, and not just as a priority. And this is what, in my view, determines a cultural transformation that is perennial.

## The importance of symbols

Symbols are the visible artifacts that communicate and reinforce the safety culture, make people around them perceive and value the presence of safety. We list below the main ones:

## 1. PPE

Personal Protective Equipment (PPE) is any device or product for individual use used by the employee, intended to protect against risks capable of threatening their safety and health.

The use of this type of equipment should only be done when it is not possible to take measures to eliminate the risks of the environment in which the activity is carried out, that is, when collective protection measures are not viable, efficient, and sufficient to mitigate the risks and do not offer complete protection against the risks of occupational accidents and/or diseases.

## 2. DDS Notebook and DDS Box

The DDS notebook can be used to put notes, important themes and subjects that will be used when applying the DDS to employees. The DDS Box is an easy-to-apply tool for management to conduct the DDS through organized, practical content that portrays the daily lives of employees in the area. Topics can be divided into Environment, Health, and Safety.

*This method allows different modes of use:*

- The leader will be able to choose one of the cards in the box and will have to read, explain, and discuss the subject defined in it.

- The leader may suggest that someone choose a card, that person will have to read, explain, and discuss the subject defined in it.

- A lottery can be held for someone to read, explain, and discuss the subject defined in the letter.

- Both the notebook and the box are artifacts that symbolize care.

### 3.  Emergency response preparedness equipment

Emergency response preparedness equipment aims to regulate the minimum requirements by the company and the government to carry out activities, mainly aiming to avoid accidents and, in the event of an accident, to mitigate its consequences on the environment and on the potentially involved public, aiming to radically reduce the possible losses of any of the production factors: natural resources, labor or technological equipment.

### 4.  Communication wall

Communication walls are important internal communication tools for the company, generally aimed at a greater number of employees. This communication vehicle is posted in places with high traffic of people and takes, in a dynamic and efficient way, content of relevance and awareness for everyone within an organization, about safety, standards, warnings, internal information, among others.

### 5.  Safety signage

Among the methods of protecting and ensuring the life, health and profession of employees who are inserted in the field of occupational safety, there is safety signage in work environments. They aim to alert employees and visitors to the site about the risks existing there, protective equipment and other essential information for safety at the site.

They were made to attract attention, with clear writing, obvious illustrations of easy and quick understanding, which helps to have a quick conception in emergency cases. Universal symbols are used in the drawings that offer standardized understanding and do not generate double meanings or any other confusion. The use of colors also has its meaning, as do familiar forms and acronyms in the workplace and around the world.

## 6.    Recognition trophies

Trophies in the safety area are intended to honor, encourage, and reward, as well as promote events, offering recognition and satisfaction for the victory of an employee, leader and/or manager.

## 7.    Training certificates

Training employees in the safety area of a company is a very important investment. Companies have always been concerned with this issue, and the offer of safety courses, lectures and workshops has been modernizing and following the evolution of technology.

Opening this possibility within companies means better preparing employees as well as increasing their productivity and engagement related to their safety and collective safety, improving the performance of managers and safety leaders, consequently achieving the company's objectives and goals.

At the end of these corporate events, it is necessary to deliver some type of certification that proves each person's participation. For this, many companies and educational institutions are looking for some training certificate model

By participating in some course, workshop or training, people expect to receive some kind of certificate. This is a very important and fundamental document in this type of situation, especially when we talk about professional qualification events.

Some details cannot be missing from the training certificate template:

☞ Logo of the company or institution offering the training;

☞ Training name;

☞ Participant's name;

☞ Date of completion;

☞ Training duration time;

☞ Certificate validation.

## 8.  Photos of the trainings

The photographs of the trainings can serve for the employee to have some memories of the day the training was carried out and also of the time and hours that were invested by them in the safety training.

In addition to creating memories of reflections and commitments assumed with the topics covered.

## 9.   Pins and stickers

One of the purposes of the pins is to be able to inform the prizes of the participants of some training or lecture and to inform the hierarchy of the person who is using the pin. They can give a different meaning to someone, as they can also be used to inform some important safety issue (significant dates such as yellow May, blue November, brigade member day, among others). The stickers can be used both for the purposes mentioned above and also to indicate a place where a specific task is being performed, and to inform about an equipment's integrity and certify its authenticity.

## 10.  Campaign printed material

The printed materials of the campaigns serve to raise awareness, important issues related to safety, as well as inform about the dates and times of the campaigns, also helping the employee with other relevant information about safety.

## 11.  Safety badges, cards, and messages

Badges and cards are used to gain access to any location and/or safety event that may occur, as well as specify the role or responsibility that the manager, leader, or employee is having at the time of using them or specify the position of the person within the company. Safety messages serve to make employees, managers, and leaders aware of its importance, as well as informing any specific safety date or internal company warnings.

## 12.  Safety performance report

The reports present the performance status, the best practices implemented and serve as a symbol of the "safety" value's transparency.

## Reporting performance is important to:

☞ Understand and communicate the progress and performance of the safety agenda considering the employee, manager, or safety leader;

☞ Correct deviations in relation to safety;

☞ Present trends in results;

☞ Celebrate and recognize good practices; and

☞ Check how resources are being used.

## 13. Accident-free days sign

The accident-free days signs are elements used to facilitate people's visual communication about the awareness of accidents and incidents, always bringing the memory of daily work safely.

## 14. Risk map

The risk map is a graphic representation of a set of factors present in the workplace, capable of causing damage to the health of employees: occupational accidents and diseases. Such factors originate in the various elements of the work process (materials, equipment, facilities, supplies and workspaces) and the form of work organization (physical arrangement, work rhythm, work method, work posture, working hours, work shifts, training, among others).

## 15. CIPA identity, brigade, and safety team

The CIPA identity and the safety team aims to prevent accidents and illnesses resulting from work, it is a team trained to face and support safety issues and serve as spokespeople, in addition to capturing, discussing and understand the open demands. The fire brigade members are the members of the fire brigade trained to prevent and fight fires, as well as provide first aid, assess risks, prepare reports, guide people, call the fire department, and other actions that aim to save the lives of everyone involved.

## 16. Uniform and helmet colors

The colors of the uniforms are intended to identify the employee's role, as well as inform the specific area they can safely access using the necessary PPE. Another important purpose of the colors of the uniforms is to bring the identification of the company in which they work, in addition to signaling if they are a newbie, pregnant, among other topics that we may want to identify.

## 17. Brigade room

The fire brigade room serves both for meetings, training, and lectures on employee safety, as well as storing the necessary materials that the team uses in their training (fire extinguisher, signboards, among others). It is a symbol of preparedness to respond to emergencies within the unit.

## 18. Work permit visible at the place of work execution

Work Permit (PT) or Permit to Work (PPT) is a prevention tool, which aims to identify and previously assess the risks when carrying out work with the potential to cause damage to people and/or property, having to be visible in the place where the work is carried out, constituting a symbol of safe work planning.

## Important information:

☞ It will limit the work to a certain equipment, area, or period of validity;

☞ All its fields must be filled in and all the sectors involved must approve it;

☞ The approved Work Permit (PT) or Permit for Work (PPT) will be held by the person performing the work, and must be presented whenever requested;

☞ In shift changes or work interruptions, PT or PPT must be revalidated;

☞ The Work Permit (PT) or Permit for Work (PPT) must be filed at the company;

☞ The Work Permit (PT) or Permit for Work (PPT) must be visible at the place where the work is performed.

## 19. Sensors, lights, and safety devices

Safety sensors and lights are one of the most important human safety devices. It is used in equipment with the main objective of monitoring an operation danger point and interrupting its operation as soon as it detects the presence of a person's finger, hand, arm, or body through the interruption of its light beams.

Safety devices are devices responsible for interrupting the electrical current, if the intensity of this current is greater than that supported by the device. The consequence of this intervention preserves the other elements that make up the circuit. The most common safety devices are fuses and circuit breakers.

All are safety symbols, they can never be disabled, blocked, jammed. It is a demonstration of care for the employees who work on these machines and equipment.

## 20. Licenses and permits

The licenses serve to give the proper authorization according to the company's internal rules, as well as governmental rules, so that some service, undertaking and/or maintenance can be started.

The permit is one of the first documents requested for the operation of a business, and the company is not authorized to start its activities until the granting of this document that certifies the suitability of the desired activity at the chosen location.

The license is a mandatory document for all types of commercial, industrial, agricultural establishments, societies, associations, institutions, and service providers, whether individuals or legal entities. It is issued by the prefectures, varying its procedure according to the legislation of each municipality.

The municipality is also responsible for monitoring compliance with these rules and may impose fines and other sanctions in the event of non-compliance. The State or Municipal Consumer Defense Agencies are also competent to carry out inspections, assessments, and application of sanctions.

Having all licenses and permits up to date sends a message that we can operate and that it is safe to operate.

## 21. Minutes of Meetings

A meeting minute is a written record intended to reproduce all events, discussions and decisions related to safety areas taken at a meeting or assembly.

## 22. Action plans

The action plan is a management tool widely used for planning and monitoring activities necessary to achieve a desired result in the safety area. The action plan allows monitoring the execution of the most important activities to achieve certain objectives and goals related to the productivity and safety of the company and the employee.

## 23. KPI's

KPI is the acronym for the term Key Performance Indicator. This indicator in the company's safety area is used to measure the performance of processes and, with this information, collaborate to achieve its safety-related objectives.

Through safety KPIs, all employees know and become involved in the corporation's mission, in order to align efforts around the strategies established by their superiors.

In this way, through the results indicated in the safety KPIs, it is possible to quantify the company's safety performance and allows employees to understand how much their activities contribute to the success of these numbers.

## 24. Physical structure

The company's physical or organizational structure in safety is the way in which the activities carried out by a company are divided, organized, and coordinated with a focus that includes the description of physical aspects (e.g.: facilities, human, financial, legal, administrative, and economic), always having as a value the safety of the company and the employee.

The good structural conditions bring the message of zeal and that there is a concern for the well-being of all workers.

## 25. Order and cleanliness

The responsibility for order and cleanliness belongs to everyone. Each employee is responsible for keeping their work environment clean and orderly, so that each piece of equipment or work tools is in its proper place, there is no dirt or materials scattered around the place.

The lack of order and cleanliness often creates problems that affect productivity and the effectiveness of operations, contribute to the relaxation of personal hygiene habits, and increase the propensity of occupational diseases and accidents at work.

When there is good order and cleanliness in the workplace, there is a more pleasant and healthy environment that reinforces the positive attitude of employees, increasing production and reducing the risk of accidents. Because of this, order and cleanliness are basic needs that form an integral part of our working environment.

## 26.  Condition of the bathrooms

The state of conservation of a company's restrooms must comply with the company's internal safety regulations and health surveillance regulations. Having this correctly, there will be a better use and productivity of employees in their functions, as well as the minimization of accidents related to employee safety. In addition to conveying a message of care for the well-being and health of workers. Here we can consider Maslow's pyramid and its importance for the self-esteem of each employee.

## 27.  Food and cafeteria

A company's cafeteria facilities must comply with the company's internal safety standards and government regulations. Food in companies must have the correct quality and follow the recommended portions for each employee. In addition to conveying a message of care for the well-being and health of workers.

Another point to consider is that the company's performance is linked to the maintenance of the employees' health. Adequate nutrition is the basis on which the employee's total physical and emotional well-being is based.

A malnourished employee has their vital functions compromised, thus being more prone to accidents and/or occupational diseases. Therefore, the correct nutrition of the employee must be extremely important for the company.

## 28. Unsafe conditions

Unsafe conditions are those situations present in the work environment that put people's physical integrity and/or health at risk. These are defects, failures, technical irregularities, and lack of safety features. Unsafe conditions send carelessness messages of no importance to the value: safety.

## 29. Vehicle status

The state of the company's vehicles must comply with all governmental safety standards established for the safety of the employee, as well as for their better use. As well as industrial facilities, vehicles and their respective state of conservation are communicating the importance of safety value.

## 30. Up-to-date maintenance seal

All maintenance seals must adhere to established government safety standards. Under no circumstances may the employee violate the equipment's safety maintenance seal or circumvent it when indicating a sector where maintenance is being carried out. Among these requirements, we highlight that the seal must have: up-to-date maintenance period, validity period of the current equipment, location, or equipment where maintenance is being carried out, person responsible for maintenance, among others.

FROM THEORY TO PRACTICE

# Inspiring Principles for a Strong Safety Culture

*To make a difference in someone's life, you do not have to be brilliant, rich, beautiful, or perfect. You just have to care.*

Mandy Rale

Over the years of studying and working with workplace safety, I've learned that the first and most important synonym for safety is care. And when it comes to care, it is important to emphasize that it exists in three different dimensions.

> The first: I take care of myself.
>
> The second: I take care of my colleagues.
>
> The third: I allow myself to be taken care of.

Any people around the world understand what care is. It is independent of customs, language, social differences, or geolocation. More than that, it does not see hierarchy, it is not paralyzed by it, it is not corrupted. And it's a basic phenomenon of human existence.

But, although it is something basic for every human being, it does not mean that it cannot be culturally learned. So much so that the way we were cared for or expressed care influences our own way of caring (Roach, 1993). Hence the importance of seeing care as something extremely alive and organic, and of humanizing safety within organizations. Care transforms safety traditionally recognized as bureaucratic into relational.

With this in mind, as we focus on creating a new safety culture, it becomes more urgent, powerful, and efficient to address the WHY instead of the WHAT. Isolated safety measures and rules do not work if we do not understand the reasons why they were created, who they are intended for, what they intend to change culturally.

In my work within companies and as a consultant, I have developed six principles that I consider fundamental in this journey to explain choices, the whys. They are the main arguments that make – or, if they don't already, should start making – the safety challenge a constant search for excellence.

To create a culture that is strong and sustainable, the work model needs to inspire, generate trust, point out the correct and truly transformative motives – those capable of uniting people on the team –, all based on the meaning of CARE.

I list the six below, in no order of importance, since for me they are all equally fundamental.

## **Principle 1:** Safety is a value

In many companies, and for many years, safety has always been seen as a priority. But priorities change over time – what is the most important thing today may not be the most important thing tomorrow. And that cannot happen. Therefore, safety needs to be a value, not a priority. And what is value to you? I like to reflect on what this means and to cite, when talking about the subject in my lectures, the example of the great navigators, explorers who traveled around the world in search of new lands and who traced their routes from the position of the stars.

Our values, as I see it, are something like this, they are our guides, our direction, our stars. They justify the way we act and the decisions and choices we make.

It is impossible to achieve excellence in safety if it is not seen as a value, as a guide for each individual and if it is not internalized in our organizations. Values are non-negotiable, they are protected and precious to us. They often carry stories, behaviors, they are the result of someone's legacy. They feed practices and provide a sense of common direction. And they set the standards that must be implemented.

It is important to remember that values can be absolute and universal, or relative. According to ethics and some philosophers, when relativized values tend to be inconstant and linked to a series of other circumstances, such as the education the person received or the social and historical context in which they live.

In these cases, what is the best way to recognize the values of a company or team, standardize and disseminate them? In my own experience, this path goes by the name of active listening. Knowing how to dialogue, understand and interpret what your employees think, what they believe, what their doubts are and what they seek is essential to adjust some values.

But when we talk about safety, we talk about an absolute value. Nobody wants to put their own life at risk, or their colleague's. It's independent of such changing circumstances. A person who values safety and care will have them wherever, whenever or with whomever they are, at work, at home, with friends or alone.

For safety practices to be followed within a company, therefore, the issue must be treated as a value, and not just as a priority. Because an accident – something that is beyond this value – is inexcusable, unacceptable, and priceless. Our lives and the lives of people under our responsibility are non-negotiable. If we can prioritize values, by the way, safety would probably be at the top of the list, since it concerns protecting lives, protecting people's physical integrity. What could be more important than that?

I believe our mission is to signify things. As long as safety is perceived as a procedure, rule, PPE, we will not be able to reach the scale of people's values. That's why it's so fundamental to signify care.

## **Principle 2:** Safety is about people

For many, and for a long time, safety meant exclusively procedures, rules, complying with speed limits, the traffic guard, the car belt or even PPE. Often, it was only important because it would be the result of an audit, of a number of hours of training. If indeed these weren't the goals, that's what the way of communicating and bringing safety into the corporation made it look like.

At what moment, exactly, do we forget that the main gear, the most valuable asset we have in our lives, in our business is PEOPLE?

Yes, safety should be about taking care of people, preventing them from not returning to their homes, to their families. Regardless of the position they occupy, I like to look at each worker and always think about who is at home waiting for them. Make sure they're important to someone.

Our safety work starts with understanding who the main customer is, who should be heard and involved in our actions.

Therefore, in this book I intend to encourage you to have a humanized view of safety, understanding that 100% of accidents are defined by the behavior of these people in the face of risk. If we want (and we do want) to avoid them, we need to empower safety people, understand their motivators, the way they make decisions. We must listen to them and bring them closer, in order to influence them to multiply safety and transform their visions through a new culture. We need to bring safety to the center of people's lives.

It is necessary to understand and know how these people get to work every day and have a genuine interest in their world is what will prevent further accidents. Or do you think that coming to work after being beaten at home, or facing a child's drug problem, or taking care of a sick mother, or knowing that a husband has lost his job does not interfere with the risks and the way this person goes to work?

We cannot ignore people, we need to be interested in them, have a genuine desire to plant the conviction of care, to prove that safety is the best choice. That is why the main tool for strengthening the safety culture is dialogue. Remember care is contagious. Whoever receives our care will take it to more people.

**Principle 3:** You are primarily responsible for your safety.

Imagine that you hail a taxi and when you get into the car you realize that the driver is drunk. What do you immediately do? Get out of the taxi, right? Because you could never trust someone under the influence of alcohol with your safety. In extreme cases like this, we almost always take the reins so we don't take risks. But what if, in the same example of the taxi, the driver was talking on the cell phone, or driving at high speed, without much prudence? Would you also ask to get out of the car? If your answer was not "yes," it should have been. Also in this situation one cannot be permissive. And for one simple reason: responsibility for our own safety is fundamentally ours, it cannot be delegated. Blaming someone after the accident does not prevent it from happening. Therefore, it is necessary to act before the damage. I like to think that we will never stumble on a big mountain, but rather on the small rocks we decide to leave along the way.

Organizations must understand and communicate this concept widely, as it translates the main safety competencies that we need to develop in people: SEE and ACT. The two go together with risk perception, but it's necessary to highlight them strongly, as people must develop attitudes that build, together, a safe environment.

It's not enough to see a dangerous situation and warn about it. Accident investigations are almost always loaded with messages like "I told you so," "I knew something could happen" – comments that do not bring back lives, much less the integrity of those who went through the episode, in general people who do not they think they are responsible for safety, that see it as a kind of entity or department that has nothing to do with them. In addition to seeing, it is necessary to act, something that still happens infrequently.

When it comes to safety, little is done and a lot is delegated. If we are the most responsible for our own safety, why is it still so common to think, within companies, that safety is something for "team x" or "department y"? Many think that the ideal world would be to have a safety technician accompanying every employee and/or feel safe only when the safety department is around.

It can't be like that. A safe environment is built daily, by everyone, it is full of "seeing and acting." More than wanting safety, it is necessary to practice safety. We all own it. Risk perception, seeing and acting are not exclusive to safety technicians and engineers, they are available to everyone who understands their role and calling. We cannot allow safety to be treated and also to behave as an Entity within the Organization.

It is necessary to understand that we are going to reach the levels of safety we demonstrate. It's that simple, a directly proportional equation.

**Principle 4:** Safety is a condition of job retention

Who are the people inside the organizations? According to consultant Simon Sinek, our teams are made up of:

**20%** of protagonists: they are those called bridge builders, they are all the time looking for ways to connect people, procedures, and actions to ensure safe behavior;

**60%** of followers: the vast majority, does not have a sense of safety purpose and ends up being guided by others who inspire them in some way;

**20%** of haters: haters are those who do not accept to do it safely, deny its importance and endanger their own life and that of others, that is, they are dangerous people.

What we want is to safely have leaders and employees who seek excellence without negotiating care as a bargaining chip. Therefore, it is essential that we ensure that the 60% of followers are always inspired by the protagonists, and never by the haters.

This brings me to a reflection on heroes, those taken as examples in the work environment. Heroes have always attracted me to their ability to, at the last second, turn the tide. They are the so-called saviors of the country. At some point, we started to wish we had these "heroes," who after all seem to have superpowers, on our teams.

But everything is not always rosy, and unfortunately our heroes often don't return home either – almost always because of their safety-hating postures. They are victims of accidents, for their myopic vision and their denial. The problem I see is that we can no longer call the haters heroes – this distorted view happens exclusively due to lack of planning, because those who put out the fire in moments of emergency end up being decorated, even if they did it in a crooked way, exposing their lives and that of others.

The reflection I want to promote here is about who, in reality, we are calling an example. Recognition is a fundamental ritual for strengthening the safety culture and, therefore, needs to be on your agenda in a regular, routine way. But who have you recognized? What kind of message have you endorsed and referenced? According to statistics, 98% of people who are involved in accidents are proactive, that is, they have legitimate desires to achieve personal and business goals, but don't think about safety at all.

Our acknowledgment messages must reinforce the importance of safe behavior, value step-by-step compliance with operational procedures. Heroes cannot be people who see themselves with superpowers, who at all times test their limits, going over procedures, looking for exceptions and, much less, provoking improvisation through the peculiar "special way" Brazilians use.

May we be emphatic and be able to recognize the heroes who work with real safety. Those who break paradigms and kick to the curb unacceptable behavior and arguments, such as "it has always been like this here," or "everyone does it this way," "we have never had an accident here," "the safe way is more expensive," " doing it safely takes longer."

May those who, on a daily basis, prove that doing it safely is the only way to do it be recognized. We need to return those who don't think or act like that to the job market, in order to save their lives, give them a chance to reflect on why they are leaving the company and allow them to go.

I remember once having to make that choice when a supervisor broke a life-saving rule at one of our factories. He had been with the company for 33 years and was working with two other third parties on our ammonia refrigeration system. The three were there in that environment without proper permission to work. Listen: in 33 years at the company, is it possible to believe that this would have been the first time he was running a service without permission? Also: do you think someone who doesn't follow safety rules or procedures does it just safely? That day, we lost a great colleague, we lost all the great wealth of knowledge he carried, but we gained something we could never buy: his life, his physical integrity – and those of his colleagues. And no patrimony of knowledge is greater than the value of a life.

It's important to remember that there are violations that are even criminal, and we need to be even more attentive and less tolerant of them. Whoever executes them cannot have a prosperous career path within organizations.

**Principle 5:** The immediate leader is responsible for the safety of his team

The dictionary definition is clear: a leader is a person with the ability to influence the ideas and actions of others. If you have followers, you need to be an example. For me, this definition goes further. A safety leader is an individual who inspires, influences, encourages, develops, and empowers safe behavior by visibly and genuinely demonstrating that safety is a value. Many leaders, however, seem to forget this.

As a leader, I like to imagine that we are all in front of a big mirror, and that our attitudes, choices, and decisions are constantly reflected in the way our teams work and act.

In the past, when it was necessary to investigate an accident, it was common to bring the injured person, witnesses, and the sector leader into the room, when possible, to accompany and contribute to the safety area, human resources and Cipa members.

The episode resembled a police interrogation and, as the search was always for the fault, the victim himself was always the main culprit. For these and other reasons, safety teams were often recognized as true sheriffs inside the factories. This model is broken. Until that happened, however, we made a lot of mistakes – in dealing with those involved, in dealings, in the establishment of action plans and, consequently, in the elaboration and implementation of the lesson learned in each accident.

Currently, the guidance and strong recommendation is that both the employee directly involved and his/her immediate leader be investigated. We follow the same approach – without a police or accusatory tone – with both because we must reinforce the concept that a team is a reflection of its leader. And investigative measures cannot be applied only to the employee. The leader also needs to understand their role as a safety leader.

## Principle 6: All accidents can be avoided

In a controlled environment, all accidents can and should be avoided. An accident is never justified. If there was an episode, it is necessary to understand what led to it, because something is wrong and must be corrected immediately.

FROM THEORY TO PRACTICE

# Leadership in Cultural Transformation: Inspiring and Sustaining Safety

*"The true measure of leadership is influence.*

*Nothing more, nothing less"*

John Maxwell

Before we begin, I would like you, the reader, to answer the following question: who is the leader for health and safety within the company where you work? If you've thought of one or two names, I'll tell you: there's something wrong there.

The definition of a leader within the safety culture says that he is an individual who inspires, influences, stimulates, develops, and empowers safe behavior within the corporation, demonstrating visibly and genuinely that safety is a fundamental value. Therefore, for the safety culture to be truly inspiring and sustainable, it is necessary that even leaders in areas other than safety take this approach for themselves. More than that, I would say that it is essential that all employees, regardless of hierarchical levels and responsibilities, see themselves as leaders when the objective is to take care of themselves and others, save lives.

And in the permanent position of leaders, it is essential to differentiate urgent actions from important ones. Urgent ones solve short-term demands. But it is the important actions that potentially guarantee the future and safety sustainability. The proper decision between one or the other is capable of promoting assertiveness in health and safety in a conscious way, with quality and solidity, demonstrating coherence and constancy.

Topics like these must be fully absorbed into a leader's culture, as core values in his decision-making.

At the same time, it is important to stress that there is no positive safety culture without leadership. However, unfortunately, it is still quite common for the vision of a leader to be restricted to a few individuals, and in a somewhat bureaucratic way. Far from being a value, safety is often just a legislative obligation, and ends up being delegated to more junior or less competent people in their leadership roles or with less autonomy. There is still a lot of thought about the "what" in safety management, when it would be necessary to always focus on the "how" and the "why."

It is paradoxical that companies increasingly perceive and discuss the role of leadership in organizational culture, but still do not do the same with regard to leaders in the safety area.

A good leader is concerned with setting goals and objectives, makes decisions about what should be done, and motivates and inspires others to do what is necessary. He fully recognizes and encourages the talents and skills of his team, and is therefore crucial in building, modifying, and maintaining the safety culture. It is not enough for a good leader to indicate paths. He performs, rather than just saying.

## Walk the Talk

There is no doubt that the vast majority of companies today view health and safety as a core value. Many even place the construction of a safety culture as a priority. However, the enormous distance that still exists between "wanting" and "considering it important" to actually "accomplish" it is clear.

A 2017 Deloitte survey in New Zealand interviewed 169 CEOs of different private and public sector companies about safety. The results showed that, when asked about accident risks, nine out of ten CEOs responded that they consider risks effectively controlled within their corporations. However, 25% said that such risks are not yet well documented (described and understood).

It is an evident proof of the distance that still exists between what is intended and what is actually achieved. Understanding what the risks are and identifying them is a crucial step in effectively monitoring them. If they are not even understood, how could they have been controlled, as 90% of CEOs believe?

The same survey found that one in five CEOs still do not have clear roles and responsibilities for risk control within the company and that more than 40% are still not personally committed to actions that promote engagement when it comes to safety.

The reality is the same in many Brazilian corporations. Seeing safety as an important value, but not experiencing it in practice, is creating a barrier to building a strong culture. Actions are worth more than words. It is necessary to act according to what is preached, to give examples, to inspire.

## Leadership factors

What, exactly, makes someone an effective safety leader and manager? Quality and productivity are fundamental, just like in other operational areas. Some research, however, points to two other very important factors in safety: care and control of employee behavior. This means that a safety leader must care for the well-being of people, relate well to the team, maintain good communication and be available, always controlling the establishment and fulfillment of goals and expected performance, without allowing motivation to drop.

# Efficient leadership

The most effective leaders are those who know how to listen before making decisions. Safety actions must be defined in agreement with the people involved, to ensure a greater degree of commitment and consequent results. By paying attention to what employees have to say, the leader also opens an important channel for dialogue to solve problems – problems that are often only noticed on a daily basis, during the practice of activities. This cooperative spirit contributes to the formation of a solid and sustainable safety culture, as demonstrated by a survey carried out in 2010 by the consultancy BST Solutions. According to their results, working on the safety agenda only with the operational frontline reduces an average of 25% of occurrences over a year. But if this agenda includes leadership, the reduction tends to rise to 40%.

## An efficient leader must:

- make worker safety and health a fundamental value within the organization;

- encourage and ensure everyone's commitment – including himself – to eliminate risks, protecting his employees and continuously seeking improvements to make the environment safer and safer;

- communicate with the team, pointing out failures and corrections to be taken;

- specify goals, deadlines, and ways to achieve them;

- give the necessary support and offer resources so that the team is able to achieve the proposed goals;

- be an example through his own actions.

My vast experience shows that the leaders who reach safety goals the fastest are those capable of inspiring their team, of reaching people's hearts so that they seek, in each of their daily actions, to do their best for the health and safety of all. For that, he needs a good dose of charisma, empathy, and enthusiasm. A safety leader's influence can never be through power, status, or authority.

## Leadership behavior

We have already seen that a safety leader must be someone who is trustworthy and who acts as an example to other employees. But what to do, practically, to achieve this? Here are 13 behaviors that undoubtedly reinforce health and safety leadership within organizations:

## 1. Speak directly

Have your own message that translates what safety means to you.

## 2. Show respect

Speak respectfully about the values, including the preservation of life and the physical integrity of everyone in the organization.

## 3. Be transparent

Talk openly about the results to encourage next steps.

## 4. Be correct

Act as an example in safe behavior, not breaking company rules, norms, and procedures.

## 5. Show loyalty

Be loyal to safety beliefs, projects, and vision of the future. Support the safety area, without running the risk of appearing "two-faced" to employees.

## 6. Deliver results

Continuously follow-up on action plans and goals, encouraging the search to surpass results.

## 7. Always improve

Critically analyze successes and errors in order to provide and cascade learning.

## 8. Confront reality

Confront limiting beliefs. When you hear "we've always done it this way and nothing ever happened" or "everybody does it this way," show that the risks are even greater when you think this way.

## 9. Clarify expectations

Always make it clear what your health and safety expectations are.

## 10. Be accountable

Conduct safety-related accountability for your team and/or unit.

## 11. Listen first

Know what the operation, the peers, the safe area have to say, before taking a position.

## 12. Keep appointments

Your word is your bond and must be kept.

## 13. Extend trust

Build a relationship of trust with teams, peers, and safety.

## Models and practical actions for leadership

"Individuals' safety behavior and attitudes are influenced by their perceptions and expectations about safety within the work environment. And the patterns of these behaviors end up being those that correspond to the priorities demonstrated, in practice, by the organization's leaders, regardless of the current safety policy." The statement by Dov Zohar, an expert in organizational psychology and professor at the Israel Institute of Technology, reinforces the role of leadership in building a safety culture.

Zohar, a great researcher on behavior and safety for more than four decades, was the first to use the term "safety climate," back in the 1980s, and it ended up evolving into something even broader, the safety culture. For the professor, leaders are only able to build a culture through active and symbolic messages to their subordinates.

Regarding the expected safety behavior within organizations, Zohar is categorical in stating that it also depends on the performance of leadership and designed a strategy that leaders should follow to improve this behavior within their organizations. According to the scholar, in addition to being efficient, his methodology is cost-efficient and simple to implement.

## It is based on three influence tactics:

a)  set daily (safety behavior) goals that depend on employee tasks (a list of DO's and DON'Ts, for example);

b)  create daily monitoring shifts to observe the team's performance and if they are approaching the goals defined at the beginning of the day;

c)  create immediate consequences for the behavior that is observed, both positive and negative.

Zohar's methodology also cites some practices to improve the safety climate in the organization, that is, the perception that people on the team have about safety.

a)  conduct monthly climate surveys, with questions that can be answered quickly (preferably with scaled numbers);

b)  send surveys to random sectors each month to create the element of surprise;

c)  climate notes will be communicated as feedback, allowing the management of each sector to compare their numbers with those of other sectors and their own, from previous months.

Another very widespread plan is that of OSHA - Occupational Safety and Health Administration -, an agency of the United States Department of Labor. For them, a 6-star leadership model for safety includes:

a)  supervision;

b)  training;

c)  accountability;

d)  adequate resources;

e)  psychosocial support;

f)  adequate communication.

For OSHA, such responsibilities must be followed by all organizational leadership, not just the leaders defined to care for health and safety. The organization also lists four different actions that it considers essential for building a solid safety culture:

**Action 1:** Communicate commitment to safety and health

A written policy should state that everyone's safety and health is one of the company's core values – just as important as quality, profitability, and customer satisfaction. Share this policy with everyone and, more than that, experience it in practice – in your decisions, when hiring suppliers, when purchasing new equipment... Be where your employees are, showing them that you act according to the rules that you preach.

**Action 2:** Set goals

Goals are key but think about creating ones that are also attainable in the short term. By establishing only long-term ones, the team feels discouraged on a day-to-day basis. Think that they should be realistic and measurable, and that they should also focus on prevention, not just final numbers of accidents and injuries.

**Action 3:** Allocate resources

Planning and a lot of dialogue with those on the shop floor is essential to provide the resources that are actually needed to prevent new accidents. Safety and health need to be part of the company's budget, with solid programs that have continuity.

**Action 4:** Track performance

Always being informed about the results and goals achieved (or not) in safety is the role of the company's leader. One way to gain everyone's trust so that even near misses are reported, so that they are avoided in the future, is by creating recognition programs and confidentiality programs.

The Zohar and OSHA models are just a few among many that exist when it comes to promoting effective leadership in building a safety culture.

In my book "Practical Guide to Leadership for Safety," I list actions of simple and immediate application – all tested and with good proven results in practice – to leverage and maintain this culture within organizations. They are divided into five major thematic blocks: Face to Face; Communication; Survey of Scenarios and Action Plans; Support and Engagement.

**Face to Face:** Eye-to-eye actions are, in my view, irreplaceable when it comes to safety leadership. The physical interaction of the leader with his team is essential for those who want to transmit values and beliefs, build relationships of trust and credibility, and achieve more engagement. Going to the factory floor, observing, and especially knowing how to listen to what workers have to say is the best way to understand fears, concerns, and insecurities. More than that, proximity and frank conversations with different individuals make so that the leader understands how they live and what their fears, concerns and insecurities are also outside the work environment - factors that undoubtedly influence the risk of accidents within the organization's workplace. Dialogues free of judgment, with respect and genuine interest show how much a leader is willing to actually ensure the well-being of his team.

**Communication:** Another fundamental item to engage people, direct actions and achieve the proposed goals in safety. Leadership participation in systemic and transparent communication with the entire team is what will build a solid and perennial culture.

Good communication must have adequate language, quality, and objectivity in the information and, above all, be well addressed to those who it intends to reach (receiver). The way and channels chosen to convey the message are very important to make it assertive. Newspaper, intranet, email, lecture, social networks, video, games, reports, there are many options to say what should be said in the search for safety.

A tool that proves to be extremely efficient – I have seen this throughout my career – is the sharing of stories and real experiences lived by the employees themselves. The so-called storytelling is a rich resource capable of creating immediate identification, touching the heart of the receiver in a special and unforgettable way.

**Scenario mapping:** Safety leadership must also be prepared to map out different scenarios and create new action plans. It doesn't matter if the leader starts from scratch or from an already established and functioning base, it is always necessary to look for other opportunities to improve the health of workers. For this, he must have a deep knowledge of the reality of his corporation - and even that of competitors - and be always informed about what has been proposed to reduce risks.

**Support:** There is no point in communicating, dialoguing, proposing actions and goals if the leader does not guarantee all the necessary tools for them to be implemented and followed. A progressive discipline policy should make clear the disciplinary measures applicable to each category of deviation or violation of safety rules, standards, and procedures. Life-saving rules must be established, functioning as a guide to safe and proactive behavior, as well as a Recognition Program for those who achieve good performance results, innovations, and exemplary practices.

All these tools must be transversal within the organization, that is, they need to cross all levels and sectors, impacting as many people as possible.

**Engagement:** It guarantees the sustainability of a safety culture. Employees must really believe in what they are doing in their routines to ensure their own well-being and that of the entire team. They need to wear the shirt, not merely follow rules, or enforced actions or a policy on the wall. They are components of health and safety management, but they are far from guaranteeing commitment. Some ways to achieve greater engagement are: promoting safety battles; arrange meetings between company employees and their families to present topics related to health and safety; elect safety ambassadors to watch over and strengthen the engagement of everyone in the organization.

A common mistake I see among many leaders I've worked with is their sole concern with employee engagement. In this search, they end up forgetting their own engagements, something crucial to the company's safety culture. There must be a commitment from absolutely all the actors involved in the process, starting with the number one in the company, going through each employee and, if possible, even reaching their families.

Engaging senior leadership is important because they have a greater ability to cascade values and influence people within the organization, mobilizing everyone for the cause. The family, on the other hand, represents the main reason why each of the employees thinks about safety, it is a strong ally in the process of cultural transformation when it comes to health and well-being.

Based on the five big themes mentioned above, here are 12 habits that every safety leader should include in their routine.

### Planning

Plan your day, so the chances of some important action not being fulfilled are zero. Spend time with your team members and help them identify issues before they arise.

### Prioritization

Providing proper importance to each thing and putting them in order, gives more to the less important. Don't look the other way, even if your schedule is tight and under pressure.

### Active listening

In each safety action carried out, it is necessary to listen to what the team has to say. Regularly discuss ways to improve safety.

### Communication

When filling out questionnaires, checklists, permissions, always be attentive and repeat the information aloud, interacting with the others present. Engage employees continuously.

### Attention

Read the instructions/recommendations/guidelines before starting an activity, even though they are already known.

## DDS

Actively participate in the DDS, educate your employees on the importance of safety.

## Observation

Throughout your workday, conduct behavioral observations of the team. Visit the team's workplace frequently and share your beliefs.

## Recognition

Show that you recognize and encourage good deeds that help build a safer work environment.

## Care

It is essential that the leader takes care of his own mental and physical health, with regular exercise and a balanced diet.

## See & Act

Always keep your risk perception sharp and practice See & Act.

**Example**

Comply with company safety procedures/rules/standards.

**Inspiration**

Always share the value of safety through genuine, personal messages. Be fair, don't allow a rule or principle to apply differently to people or groups.

## O'Neil's Example and Alcoa's Success (A story every leader should Know)

In October 1987, Paul O'Neill made his first speech as president of Alcoa, one of the largest aluminum producing companies in the world. The company was not having a good period and investors were anxious about the changes that the new CEO would bring about. But O'Neill, to everyone's surprise, didn't talk about profit margins or possible new markets. "I want to talk about the safety of our workers," he said. "Every year, countless Alcoa employees suffer accidents that injure them to the point of taking them away from work for a day. Our safety records are better than the North American average, especially considering that here you work with metals at temperatures of 1500 degrees and machines capable of ripping the arms of those who operate them. But these records are not good enough. I intend to make Alcoa the safest company in America. My goal is to reach the zero injured mark."

O'Neill encountered resistance from the company's Board of Directors in his endeavor, but he did not let up. When he finally retired as president in 2000 to become the United States Treasury Secretary, the company's market value had increased from $ 3 billion in 1986 to $ 27.53 billion. And its net income had quintupled, from $ 200 million to $ 1.484 billion.

The company's financial growth was linked to improved safety for Alcoa workers. O'Neill attacked the company's key habits and saw the changes that took place from there take over the entire corporation - until it became one of the safest in America.

FROM THEORY TO PRACTICE.

# Behavioral Triggers

*Culture is what remains after forgetting everything that has been learned.*

André Maurois

Trigger. According to the dictionary, everything that, like the trigger of a firearm, triggers some change, reaction, process, etc. Here we consider the behavioral triggers that lead to a low perception of risk, inappropriate actions and, consequently, accidents. If we want to preserve lives and avoid episodes that somehow threaten the safety of workers, it is necessary to map out and get to know in depth all the possible situations that lead to them, that trigger them.

When it comes to human behavior, the list of likely triggers for risky behavior is huge and constantly changing. Some of them happen at the time of an inappropriate action, which makes it easier to identify them. Like when an employee has just left the boss' office, after a tough conversation, and inattentively gets hurt on a machine. Others, however, are related to the worker's personal life, away from the workplace, or may be something he prefers to keep private. These, of course, are more complicated to map. But still, they must be addressed.

And how to approach and incorporate behavioral triggers into safety routines? The continuous effort within organizations so that the safety culture is strengthened, day by day, is constant, optimizing processes and innovating in the approach, which must be as integrated as possible into operational routines. When addressed and treated, these triggers become the materialization of See & Act, through a heightened perception of risks, and build an accident-free environment, where active care is valued.

With a simple, clear, and innovative approach, humanized safety is gaining more and more space and must be worked in a transversal way, moving beyond safety rituals, leaving the safety entity aside and being on everyone's lips, throughout the operation.

Walking around the factory, talking to workers, promoting individualized chats, assessments, and feedback, promoting quality active listening, is essential for knowing what triggers risk actions.

Behavioral triggers represent emotions, feelings, and thoughts; therefore, they are unstable and can be leveraged by four different groups of situations: social, cognitive, psychological, and physiological. Get to know each of them below, listed with 102 behavioral triggers that I have already witnessed throughout my work in health and safety, within different companies. It is important to remember that there are many more than just these 102 triggers, there are many new ones emerging every day – therefore, an organization and its safety leadership must always be prepared to identify them.

## Social

The SOCIAL group triggers are those related to the work climate, the feedback received, each one's performance, the relationship with colleagues, management, negotiations regarding goals, etc.

1. Does anyone have financial debts with a co-worker?
2. Does anyone feel intimidated to do some kind of work (knowledge vs risk criticality)?
3. My boss can't stand being given a right to refuse.
4. Do we have a new leader for the area today?
5. Did any unusual activities take place in our area today?
6. Do we have close relatives working on the same work fronts?
7. Is there a game in progress? Does anyone want to scare someone?
8. Did anyone return from sick leave today?
9. Did any machine/equipment/tool breakage occur in the previous shift?

10.  Are any releases/work at heights planned for today?

11.  Were any safety systems unavailable today?

12.  Did any accidents/incidents happen during the previous shift?

13.  Do we have any plant and/or system shutdowns planned for today?

14.  When have you seen someone breaking the golden rules and/or saving lives?

15.  Is anyone afraid to use the right of refusal?

16.  Do you feel confident and cared for when you use the right of refusal?

17.  Is there anyone who is currently expecting a response and/or feedback from you?

18.  Are you a sponsor to any new employees?

19.  Did the leadership embarrass someone on the team at the last reunion and/or meeting?

20.  Did we make any technical adjustments here over the weekend?

21.  Is anyone receiving a demand for work that conflicts with other areas?

22.  Is anyone experiencing conflicts with co-workers?

23.  Does anyone have a function deviation?

24.  Does anyone not have clear responsibilities?

25.  Is someone basing their behavior on the habits of former employees?

26.  Did we communicate at the last shift change?

27.  Is anyone having a problem with someone from another shift?

28.  Do we have someone taking on a new position/responsibility?

29.  Does anyone think they won't hit the target this month?

## Psychological

The PSYCHOLOGICAL ones concern the mental health of workers, which is directly influenced by family relationships, affective relationships (dating/ marriage), self-esteem, friendships, etc.

30.   Is anyone going through a divorce process?

31.   Has anyone been involved in an accident/incident over the weekend?

32.   Does anyone have a post-work test today?

33.   Does anyone have a birthday today?

34.   Is anyone experiencing any recent trauma?

35.   Is anyone experiencing depression at home?

36.   Does anyone experience domestic violence?

37.   How many support their homes on their own?

38.   Is anyone experiencing external threats (loan sharks, thieves)?

39.   Is anyone treating someone who is terminally ill?

40.   Has anyone buried a loved one recently?

41.   Has anyone experienced a suicide situation in the family or among friends recently?

42.   Did anyone argue with the family today before leaving home?

43.   Is anyone behind on rent (or some other bill)?

44.   Is anyone going on vacation today?

45.   Today is Friday and we have a holiday, has anyone planned to go out with their family?

46.   What time do you get home each day?

47.   Who is waiting for you at home?

48.   Is anyone waiting for a response regarding a medical diagnosis?

49.   Does anyone have a sick child at home?

50.   Did anyone spend the night in the hospital waiting to hear from someone else?

51.   Does anyone have a loved one hospitalized?

52. Does anyone have their wedding scheduled for an early date?

53. Is anyone experiencing anxiety attacks?

54. Does anyone have a fear/phobia to carry out a specific activity?

55. Has anyone recently been diagnosed with an infectious or terminal illness?

## Physiological

PHYSIOLOGICAL triggers are those linked to the physical health of the worker. Currently, physiological concerns revolve around: sleep and fatigue.

56. Is anyone doubling/working overtime today?

57. Did anyone get an injection yesterday?

58. Did anyone have an asthma attack last night?

59. Did anyone get medical attention yesterday?

60. Is everyone feeling well to work?

61. Does anyone have diabetes or heart disease?

62. Is anyone taking prescription medication?

63. Is anyone experiencing pain as a result of the activity performed?

64. Is anyone pregnant?

65. Is anyone on pain medication today?

66. Has anyone ever had an accident/incident at work?

67. Did anyone go out to celebrate and drink too much yesterday?

68. Is anyone feeling unwell today?

69. Does anyone have any sensitivity or allergy to substances handled during the activity?

70. Does anyone have any movement restrictions in their body?

71. Is anyone working in an un-ergonomic condition?

72. Is anyone over 50?

73. Is someone carrying out an activity with an inadequate number of people?

74. Is anyone working with a fever?

75. Does anyone suffer from insomnia?

76. Is anyone having trouble eating properly?

77. Is anyone having trouble adapting to a shift?

78. Does anyone carry out an activity during the period of vulnerability (extreme fatigue), at 4 am and/or 2 pm?

79. Does anyone have an inappropriate size, height, weight, or strength for their activity?

80. Is someone pushing their body to the physical limit due to an activity?

## Cognitive

COGNITIVE ones have to do with the knowledge that the worker has about the activity he performs, how much he recognizes risks, dangers, and control measures.

81. Do you know what to do in an emergency?

82. Is anyone having trouble getting off autopilot?

83. Did anyone observe a risky situation the day before?

84. Are all new employees' sponsors present today?

85. Does everyone know the brigade members?

86. Are any safety/emergency systems unavailable today?

87. Are there any activities on this shift without a procedure?

88. Has there been any change in the performance pattern of any activity?

89. What does active care mean?

90. Think of 5 main failures that can cause accidents in your activities.

91. Can you define in five steps the activities that need to be carried out today?

92. Is there someone by your side who has drastically changed their behavior in the last few days?

93. Who remembers the last accident lesson learned in our unit?

94.  Are there any new employees joining the team?

95.  Did anyone return from vacation today?

96.  Who is overdue/behind in training?

97.  Who remembers the contents of the last DDS?

98.  Has anyone participated in a DDS on the risks of this activity?

99.  Have you received training for this activity?

100. How long ago did you last perform this task?

101. Does anyone have difficulty understanding directions from their immediate leader?

102. Have you performed this task before?

## Real stories: triggers in practice

Throughout my career I have witnessed – either by eye or as a listener – countless stories with sad endings, all the result of a lack of risk perception, the value of safety, a strong and efficient culture. I share some here because I believe in the power they have to transform, to raise awareness, to lead to effective changes.

**Sugarcane plantation:** "My cousin and I were working on the sugarcane harvest and we were going behind the tractor, on the third work shift, picking up the cane that was on the ground with our hands. This cousin of mine had just come from the Northeast, to earn some extra money for his wedding. We walked about 20 minutes and suddenly I missed him. I stopped and started asking everyone if they had seen him, asked them to check in the bathroom, and nothing. So I decided to make my way back, and I found his body, which had been crushed by the tractor. It was horrible, it's hard to talk about it to this day."

This case mixes several behavioral triggers that resulted in the worker's death. There is the cognitive aspect, because he had never worked on it, he had just arrived from the Northeast and, therefore, had no experience. There's the physiological trigger, because they were on the third shift, which requires a different metabolic rate. There is the psychological aspect, if we take into account that the boy came from the Northeast to save money to be able to get married. And there is the social one, thinking about the work team's organization, which took a long time to account for his disappearance.

**Water tank:** "My brother and I worked in the same factory, he was just married and had a 1-month-old son. I had a position in production and he had one in maintenance. One day the work ended and I went home. Hours later my sister-in-law called me asking about him, who hadn't shown up yet. I called the factory and no one had seen him. We spent a sleepless night waiting for him. The next day, as they knew he was going to clean the company's water tank, they decided to check the place. He had fallen, probably because he was not wearing a safety belt, and had drowned. To this day, his son, my nephew, asks me what he was like."

**The trigger there was cognitive:** the employee failed to comply with a safety rule, which was the use of safety belts.

**Scaffolding:** "On my brother's first day at work, at his first job, he was asked to dismantle a scaffold. He started to dismantle it and, without realizing it, grabbed the electrical wiring. He stayed there, clinging to the wiring carrying the discharge, until everyone rushed to turn off the power. They succeeded and he then let go of the wiring, lost his balance, and fell off the 10-meter-high scaffold. He didn't come home.

Sad story with a cognitive trigger – it was his first job, first day at work, and the boy was probably not well prepared for the task – and a psychological trigger, due to the pressure of having to handle the task, on his first day.

**Dipyrone:** "My father went to operate his left little toe, a simple surgery that had been scheduled for some time. My father was allergic to dipyrone, and this information was always in his records. The surgery went well, but the next day he complained of pain and the nurse applied dipyrone, which caused anaphylactic shock. He died instantly. During the process we opened, the nurse cried a lot and revealed that that day she had left her little daughter sick at home. And that she was worried about her."

This story is an example of when one of our mistakes impacts someone else's life. It's something more common than we would like, unfortunately. Here, the psychological trigger was strong since the nurse left her sick daughter at home.

**Stream:** "My mother asked me to stay at home that day, with my sister, I don't even remember why. I was 12 years old and I was watching TV, keeping an eye on my sister, who was younger, through the window. There was a stream next to it and she was washing clothes in the water. As my sister suffered from some disorders, every now and then I would look outside to see if everything was ok. Until one time I looked and didn't see her. I left desperately, but it was too late. She had felt sick, had fallen into the water, and drowned."

This story shows how the experiences you go through throughout your life can influence the way you care for people. This employee was extremely zealous of himself and his team, because without a doubt his vision of safety was impacted by this episode. The trigger for her accident was physiological, and he never forgot it.

**Lack of PPE:** "I have not heard this story from third parties, I have witnessed it. It was the first fatality I faced in my career. There was a third party who provided service to us, he did various maintenance work. His price was unbeatable and he had good knowledge of our processes and facilities. We had a project to change the ceiling tiles and we had a budget with some companies, and his won the competition. They worked on Sundays because we needed to stop the factory for that. On one of the Sundays, with the brigade accompanying the work, all of the team went out to lunch, and left the area isolated. He arrived to supervise everything, went in because he was known at the concierge, and decided to go upstairs to see how things were going. But he went up without any PPE, lost his balance, fell, and died in our unit. Even with all the experience he had, he made a mistake."

**There are several triggers in this case:** his cognitive one, because he trusted himself too much and stopped using safety equipment, and his psychological one, because he wanted his company to deliver a good service and keep getting called. There was also, on the part of the team, another cognitive trigger, which was the lack of organization to ensure safety. Isolating the area wasn't enough, you had to have someone there full-time.

**A life for 50 dolares:** "Donatan came from Minas Gerais and was living off of odd jobs in São Paulo. He was 24 years old. A company that knew it would not be able to inflate or service the tires of off-road equipment at the unit itself decided to break the rule. But they had a tire repairman who would go with his tire repair shop inside the car and come into the factory from time to time to solve problems. That day, the tire repairman was busy and quartered Donotan. For $50.00, Donatan went, not knowing the procedure, not knowing the difference between a passenger car tire and an equipment tire. When he filled the first one and the tire hit the wall, it disintegrated and flew right into the boy's head. After nine days in the hospital, Donatan passed away. The tire repairman ran away."

This is another story full of triggers. On Donatan's side, there was the cognitive aspect - the lack of knowledge of the task and the risks -, there was the psychological aspect of someone who lived off of odd jobs and needed money to live, there was the social aspect on the part of the company, which hired someone who was not qualified, and the psychological one due to the urgency of having good tires working perfectly. All triggers from all actors in this event combined culminated in Donatan's death.

Finally, I want us not to forget that behavioral triggers will always explain the reason for the behavioral deviation, the unsafe act, and the accident.

Triggers can also be highlighted within the incident investigation process. When we work with investigative methods, of immediate, basic and root cause of the occurrence, the immediate cause is the behavioral deviation itself. The basic cause is the triggers that can act in combination. And the root cause is linked to cultural aspects and the immediate leader.

In this sense, I think it is urgent to transform triggers into non-superficial routine approaches that allow our employees to feel genuinely cared for.

Our rituals, symbols and heroes constitute three important platforms for stimulating and exploring behavioral triggers.

# Cultural Transformation and the Niosh Model of Control Layers to Prevent and Anticipate Risks

*People don't stumble on mountains, but on small rocks along the way..*

Andreza Araújo

## Risks and risk perception

Being able to create a safety culture strong enough to prevent and anticipate risks, preventing accidents from happening, is (or should be) every organization's goal. In this chapter we will discuss some models that prioritize risk control in order to make it more effective. But before we get there, let's define what exactly risk is.

Risk combines the probability or possibility of the occurrence of something specific with its consequences – they are always negative, synonymous with danger, failure, harm, threat.

As a bad thing, risk is assumed to be something to be avoided. Always. But in practice, this is not what happens. If humanity knows that driving after drinking alcohol increases the risk of accidents, why are so many young people still getting behind the wheel? If sex without a condom increases the risk of contracting a sexually transmitted disease, why do so many people still dispense with it? The answer to both questions: because, although the risk exists, the person's perception of it leads to wrong (and risky) decisions and actions. "It won't happen to me," "I hardly drank anything," "I always do this and nothing ever happened to me," "the chances are small"... There are many excuses when risk perception comes into question.

Risk perception refers to the ability that each individual has to identify risks – whether in affective life, at home, in traffic or in the work environment. Therefore, this perception needs to be worked on within organizations, daily, so that new accidents are avoided. Controlling risks is essential, but if this control is not accompanied by intense exercise to sharpen perception, damage will continue to occur.

But how to invest in this perception? First of all, it is necessary to be aware that it is something extremely individual and that it is influenced by various factors, such as attention, knowledge and previous experience, culture, level of physical and emotional health, excess self-confidence. A bad night's sleep can be enough for it to get scratched. And, when this happens, the real risk tends to be very different from the one perceived by the individual, which can lead to accidents of different proportions.

## Factors that influence risk perception

There are many factors that can influence human beings in their risk perception. The Campbell Institute, of the National Safety Council of the United States, divides them into three different categories: the macro, meso and micro levels. The macro-level ones are structural or institutional ones, the meso-level ones are those that involve workers as a team and the micro-level ones are the individual and psychological ones.

**Macro-level factors:** Here the culture and attitude of the organization's safety leadership have strong influences on each worker's perception of risk. When leaders show great concern about safety, this positively contaminates others, and perception increases, decreasing risky behavior and chances of accidents. The trust that the worker has in the organization also counts as a macro factor. If he thinks that the company he works for does not value the safety of its employees, if he imagines that nothing will be done if something happens to the life of a worker, he tends to act without paying much attention to the risks around him. The logic is similar to traffic tickets. While they are not actually practiced, it is more difficult for drivers to follow safety regulations, such as wearing a seat belt or respecting the speed limit.

**Meso-level factors:** are those that are influenced by peers or the rest of the team. When making a decision and adopting a risky behavior, a person often follows a colleague's or the majority's guide more than their own. In addition to decreasing the individual's degree of risk perception, the influence of others almost always leads to fear of judgment. Can you imagine being the only one to use all safety equipment in an environment where no one else does? It takes courage to "row against the tide."

**Micro-level factors:** in these cases, the perception of risk is defined by the level of prior knowledge that each individual has. It is assumed that the less information a worker has about a procedure, the more care he takes when performing it, the less risk he will want to take. Of course, we are not saying here that, for this reason, workers should have less information. Obviously they need to be well trained and knowledgeable but paying extra attention to self-confidence does not increase the chances of wrong decisions that increase risks and lead to accidents.

Optimism is another very common micro-level factor in organizations. It's the good old "this will never happen to me/us."

# NIOSH Risk Control Model

The lightbulb in the living room lamp burned out and needs to be replaced. Even though a simple and common task, indoors, requires some safety measures. The risks of being shocked are small, but they exist – therefore, it is recommended to always unplug the part, reducing the possibility of electrical discharge to zero.

Completely eliminating risks is the most efficient action in the search for safety. It is known, however, that elimination is not always possible in industrial processes. Therefore, to reduce accidents in work environments and protect workers, the National Institute for Occupational Safety and Health (NIOSH), in the United States, created a model that ranks the actions that can (and should) be taken for more efficient risk control. They are listed from the most to the least efficient and protective. The idea is that, following the hierarchy proposed in this inverted pyramid reproduced below, corporations can implement safer systems, reducing the risks of occurrences.

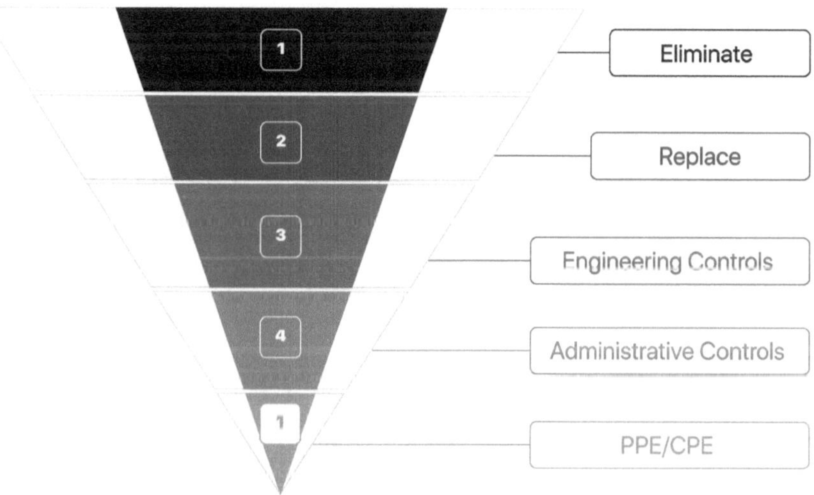


Figure 10: Niosh model to prioritize actions to control risks.

Control layers were defined by:

**Eliminate:** Physically remove the hazard. It can be a machine, a product or even a process step. The most efficient action in the hierarchy in general is also the most difficult to implement, since there may be resistance to changing old processes that workers are used to. Not to mention the high costs that elimination can involve.

**Replace:** Replace the hazard with something that is safe (or reduces risk). This step also involves a certain degree of difficulty, for the same reasons as elimination.

**Engineering controls:** keep people away from danger. By creating engineering projects capable of reducing risks, such as ventilation systems, gates, machinery shutdown mechanisms, overall, more sustainable safety is created, which is maintained in the long term.

**Administrative controls:** change the way people work. It is a necessary control, but not always efficient because the reduction of risks will still depend on the way workers act. Of course, it is necessary to carry out training, create warning boards and preliminary risk analysis (APR), the stimuli to strengthen this layer must be constant.

**PPE/CPE:** Protect workers with PPE/CPE. Personal and collective protective equipment assumes that risks exist, but that they cannot be completely eliminated. That's why it's last on the hierarchy list and tends to be used when none of them has been fully effective.

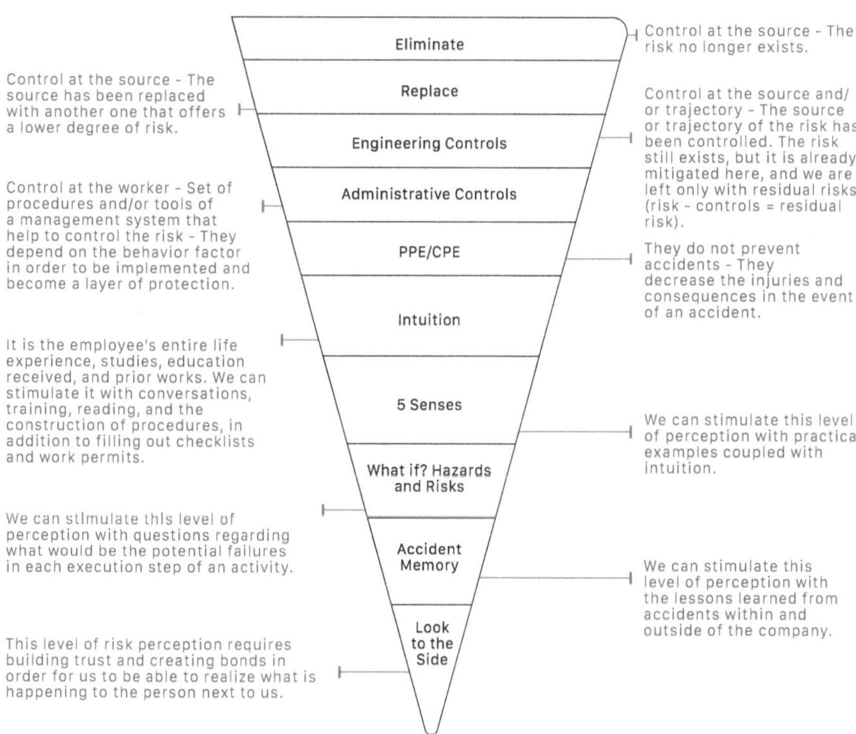


Figure 11: Modelo Niosh adicionando os elementos de Percepção de Risco.

The NIOSH model is important and should be considered. But my experience shows that, many times, the actions of elimination, replacement, engineering, and administrative controls tend not to be developed by the individual user, the one who is in fact most exposed to the risk, which means that he may not have participated in the control process as a whole and have been involved in this model exclusively by receiving PPE. And one cannot limit the look at the worker only to the PPE level.

In addition, the individual perception of risks, which we have already discussed, comes into play. Therefore, when looking for permanent and real changes with regard to safety, it is essential to enhance the safety culture in each individual, which translates into the habits and choices of everyone on the team. From the moment we hand over PPE to an employee, we must continue to encourage them to behave safely – or things will go back to the way they were before.

Here I propose a new model of protection layers, a hierarchy capable of providing stimulus for the development of the core competencies of a strong safety culture. In it, the perception of risk materializes the See & Act.

In safety, risk perception, as we have seen, is the act of becoming aware of a danger through the human attributes of "thinking & feeling," it is thinking about the possibility of getting sick or having an accident as a result of exposure to such a risk. It's being a few steps away from an accident and being able to think ahead, understand what the controls and needs are so that this activity can be carried out safely, and avoid it.

Different analyzes over the years, including one by the Department of Psychology at the University of Sheffield (1), have highlighted the role that risk perception plays in making safe decisions and have shown that when risk perception is changed, the behavior of people in the face of safety almost always changes as well. So the culture changes.

But how, exactly, is risk perception formed? There are those who see risk in almost everything, with a very high perception, and there are those who overlook any and all signs of risk that they see ahead. Some other characteristics of each individual have a certain influence on this look – even if the person is optimistic or pessimistic in life, it can interfere.

Something that also often influences is how attentive to numbers a person is. Imagine a specific machine that presents great risks, with which there have been dozens of less serious episodes, but which serve as a warning to form the perception of risk about it. Someone more attentive will think: "wow, if 37 people have almost had an accident there, the chances of me having an accident are also greater," which makes a little light go on in your head. To the most distracted, the risk will go by unharmed.

Personal experiences also influences our perception of risk. If I or someone I know has had an accident in a certain situation, I tend to be more careful when going through the same process. The same goes for training, for example. If I received specific information about a particular machine or product and was previously alerted to the risks, my perception of them ends up being influenced.

When we talk about risk perception, we are talking about human behavior. And it is known that affective and external factors and contexts are capable of dictating an individual's behavior. If I arrive at work insecure and disappointed by any situation I went through outside, with greater difficulty to control my emotions, the tendency is that, on that day, my perception of risk is lower. That's why, based on my experience, I believe so much in risk perception as something fundamental at work to build a strong and sustainable safety culture within corporations.

See & Act is the materialization of action, it's a step forward in perception, eliminating, correcting, and interrupting activities that are being planned and/or carried out without the necessary safety conditions.

I remember a leader I had in the past, who always challenged us in the face of the lack of See & Act in his team. He always teased us with the same phrase: "Doing nothing is not an option."

A more concrete example that I bring with me was one of the many accidents I have witnessed throughout my career. It happened to a worker named Ana Paula, then 25 years old, head of a family and mother of a little girl of 3. Ana Paula was a new employee, working with production planning in the corporate department and one day she received a demand to visit one of the company's manufacturing units, in the countryside of São Paulo. A good employee, the next day Ana Paula left in the middle of the night, when she arrived at the unit the day hadn't even dawned yet. As she had never been there, she stopped just before to ask what the main entrance to the factory was. She was advised to go to a technical services concierge, which had specific opening hours. As it was still dark and there was no one at the entrance, Ana Paula decided to stop her car inside. The rail gate was not padlocked, and when she tried to push the gate open to secure her car, it fell on her body. The gate had been broken for two years, and that wasn't even the lobby where the employee was really supposed to be. Ana Paula did not survive the trauma.

In this tragic case, a paper with the inscription "broken gate" could have saved a life. But nobody thought of it. Nobody imagined that a person would try to access the factory through that deactivated ordinance. And no one imagined that a broken gate could kill. The lack of risk perception, our autopilot and the non-execution of See & Act have taken the lives of many people. As well as the lack of empowerment of people in safety and the idea that someone will always take care of it for you.

Another big problem within companies is conditioning the perception of risk to a capex or extra investment. I am against permanent improvisation, and whenever there is a risk it must be definitively eliminated or controlled, not just alerted. But "doing nothing is not an option," we need to act – especially when our perception of risk is sharp! Yes, the gate should have been fixed. But if it wasn't, a poster about the damage had to have been put up.

Risk perception is a growing one that must be constantly stimulated in each employee. With this in mind, I propose a risk control hierarchy model, created based on my experience in the safety area, which goes beyond that of NIOSH. It includes the following steps:

Figure 12: Risk Perception Ladder Model

**Intuition:** it reflects the experience acquired by the worker throughout life, everything that is learned outside the work environment. Learning assumptions and life experiences work as triggers in specific situations and cause doubt in continuing with some activity when we see some risk in it.

**The five senses:** smell, sight, taste, touch, and hearing can also help in identifying a non-standard condition. It is worth promoting simulations and stimuli that promote the use of these five senses.

**Dangers and Risks (What if?):** it is necessary to create in people the habit of questioning, creating possible assumptions, thinking about failures and their consequences. Involving employees in the development of qualitative and quantitative risk analysis tools can favor understanding of existing hazards and risks and all control and preventive measures. In each industry or type of activity, we are aware of the existence of deadly hazards, hence the importance of knowing the tasks, their hazards, risks, and the respective control measures so that there are no accidents. Before applying quantitative methods that measure frequency, probability, and severity, it is recommended to use qualitative methods, such as the "What if?" methodology, which helps people to think about possible failures and their respective consequences. We need to train our employees to look for faults, because if they and their consequences are understood, we will have a different attitude towards exposure to dangers and risks. We must reflect on lions and sharks in our daily lives, so as not to turn them into kittens and dolphins.

**Memory of previous accidents:** almost every individual has some memory of personal events, of some past history, of an old accident that he witnessed or even experienced. Preserving, retelling, and keeping these stories alive helps people see themselves in risky situations. Furthermore, we can say that the memory of the other can help activate my own memory. One of the most emblematic cases I have ever witnessed about accident memory was during a visit to a factory in Brazil, a cement plant, and when we went by the cyclone tower there was a memorial built there. It contained three engraved names and a date – I asked what it was about and I heard the answer: "Three employees lost their lives here, on this very day, in an accident that we never want, nor can we, forget." I was surprised to see that, but at the same time, extremely impacted by the power of that tool. Safety, I have no doubt, can only be built if there is the courage to face it and the ability to expose one's own vulnerability.

**Looking away:** the human being needs to be seen and recognized as an individual, and not just as a team. You have to learn to look the other way and avoid questions, automatic answers, and prejudgments. Fostering the practice of empathy is essential to captivate and consequently establish bonds of trust between all employees.

FROM THEORY TO PRACTICE

# The Practice of Care and the Power of Empathy

*Knowledge is meant to charm people, not to humiliate them.*

Mario Sergio Cortella

**The name says it all:** behavioral observation. This look around paying attention to HOW people work is a lifesaving tool that I consider an excellent thermometer for measuring the unit's risk perception, culture, and safety climate. It's already widely used in the market when seeking to change risk behaviors and maintain safe behaviors, but it's not always carried out in the best way.

I chose to dedicate a space to this tool in this chapter, as I recognize its catalytic potential by transforming behaviors, re-editing beliefs and directly contributing to the safety culture.

It's necessary to encourage behavioral observation and show the entire team that safety must be seen as a non-negotiable value, and that it depends on everybody.

It's unfortunate when, during the investigation of an accident, it's discovered that something was wrong and that someone knew about it, but decided to look the other way, believing that everything would be all right, that nothing bad would happen. Risks are present everywhere, at all times, and all safety procedures have a reason to exist. Closing your eyes to the behavior of others – and sometimes even your own – is to exonerate yourself from the responsibility of saving a life!

And why is human behavior so important when it comes to safety? One of the fathers of Behavior Based Safety, American psychologist Scott Geller, a professor at Virginia Tech, proposes that with simple but efficient observation techniques, employees observe each other, identifying safe and risk behaviors. Each observer must then provide individual, constructive, and empathetic feedback in order to reinforce safe behaviors and propose alternatives to risky ones. This should be done without pointing the finger at or publicly blaming those observed.

These feedbacks are important and powerful but not enough. We know that an individual's behavior, whether safe or risky, is directly and indirectly influenced by the work environment and individual issues. Therefore, it is necessary that we look for the systemic causes of the observed risk behaviors. If they're not wearing masks or gloves, let's ask ourselves why. It's very possible that the problem is merely a symptom of something deeper within the organization.

And in this dive in search of the most rooted causes, an important keyword must be kept in mind: empathy. It's necessary, in fact, to have it as a guide from the beginning of the process, even before the behavioral observation actually begins.

Next, we will present 17 pieces of advice that guide my behavioral observation.

# 17 Advices
## for an appropriate behavioral approach

### 1.  OPA – Observe, Plan, Act!

Observe people, the layout, distribution, people, interventions and isolations, everything that is happening in the area or department you choose to address. Plan by choosing the person and/or group you will approach and act. The best way to approach the person is by making a hand signal or entering their field of vision. It's important never to put yourself in a risky area to approach a person, never to frighten them. And remember that the process of "Observing" is not about identifying what is going well or badly; it's, as the name implies, only an "Observing" of facts and behaviors, defining whether they are safe or unsafe, in order to determine, later on, the way in which we will approach the worker.

Not passing judgment on the situation allows us to approach the worker in a neutral moment, which will make him understand why the person is behaving in that particular way, and, with that, influence them in a better way so that, at the same time, at the end of the behavioral observation, they develop a personal commitment to safety. It's necessary to understand that when we observe an activity and define whether it's "good or bad," we are making a judgment and this does not allow us to hear and understand why the worker made the decision to behave in that way.

If we observe that the observed person or persons are working safely, it's necessary to acknowledge them for this and immediately thank them for their behavior; with this we will reinforce the behavior that we expect the worker to have and show that we have the ability, as leaders, to observe the activity done correctly.

## 2.  Can we talk?

Start by introducing yourself and explaining what you are doing, shake hands with the person approached, and call them by name. Eye to eye is the shortest way to get to the heart. The conversation needs to evolve from the Why to the How and What.

Ask for safety instructions, ask for the safest place to stay while observing, ask what the person is doing, and don't allow them to continue working while you're talking.

## 3.  Start by conncecting with something positive

The conversation should be guided by your observation and capture the observed person's perception of safety. Break the ice during the conversation, reinforce something positive about work, family, football team, etc. Another way to connect with the observed is to look for commonalities. In fact, this is my favorite way.

## 4.  Active and curious listening

Safety culture is a process and therefore takes time to develop and requires a collective effort to implement. The first step for an efficient and sustainable culture to exist is to capture and deeply understand the beliefs and values of the people within a corporation. There is no other way to successfully engage employees without active listening.

Listening is a fundamental skill for any and every leader. But not just any kind of listening. You need to listen actively, that is, show that you are interested in people's thoughts and opinions. It's necessary, for everyone, together, to build a sense of pride in belonging to the same organization, a responsibility to care, act and ensure safety.

In active listening, one hears with the ears and eyes, in a quest to connect with thoughts, feelings, anxieties and beliefs. More than that, you suspend your own judgments and beliefs, undress your own thoughts and anxieties to really engage with the interlocutor. It's observed without halters.

For many leaders, such an attitude is still seen as a "disrespect for hierarchy." But, you see, it's not about disrespecting, it's about letting go of it. When the leader proposes to listen in order to measure and recreate a certain current culture, he must forget about hierarchical levels. The voice of everyone on the team is important and can – actually, should – be heard.

It's also important to emphasize that listening goes far beyond giving someone your full attention, in silence. Especially within organizations, precisely because of hierarchy, people do not always say what they really want to say. It's necessary to pay attention to many other signals such as body language, mood, facial expressions, and behavioral tendencies, both individual and collective.

Listening needs to be a full-time job. And from the entire team, regardless of position. It requires showing genuine interest, being present, making eye contact and prioritizing the person being observed to speak.

## 5.   Mirroring

This is an important technique for creating empathy and identification. Study the person during the conversation and notice their breathing rhythm and then mirror their facial expressions, their tone of voice and your body positions to theirs. Bring into the conversation slang and accents that the observed presents. But be careful: don't imitate the person in an explicit way, don't act like a robot. And never scream, even if they do.

## 6.   Powerful silence

One of the basic tools in observation is the silence phase, when we must let the worker speak and define how and why a change to safer behavior will begin. You need to help them identify what would help them stay alert to the risks and how they could be aware at all times of what might happen. This silence, in general, includes allowing between two and three seconds between each question the leader asks, making room for the unexpected.

## 7.   Ask, ask, ask

The question seems very simple and logical. However, the challenge at this stage is to ask powerful non-judgmental questions that make the worker reflect on what they are doing and the consequences of their actions without the supervisor warning them. It's not about talking about whether what's being done is good or bad but promoting self-observation.

Questions should be aimed at challenging what people believe (it won't happen to me, we always do it that way, I'm an expert, we're late, etc.), they should be as simple and short as possible to make the person reflect, self-analyze the situation, and verbalize what is happening.

## Some suggested questions to approach during observation are:

- Why do you think I came here to take care of you?

- Tell me, what are you doing?

- What do you think of this situation?

- Why are you working this way?

- Is there a safer way to carry out this activity?

- What do you think we should do to make your daily life safer?

- What do you need?

- Why would you work more safely?

- Who did you leave at home today?

- What dreams do you still want to fulfill?

## 8. Be prepared for diversity and inclusion

You may find someone with a disability and different gender identities and sexual orientations, which should not stop you from conducting behavioral observation. Include all people and ask the observed leader for help if needed. Never stop an approach due to the difficulty of inclusion or judging the observed person for their personal realities.

## 9.  Strengthen current tools

Explain what behavioral observation and other unsafe conditions reporting tools are, as well as the lifesaving tools used by the organization. Never complain about the company's lack of safety tools or badmouth the safety department or its leadership.

## 10.  Ask about other safety concerns, give room

Check if there is any situation that bothers and worries the observed due to its potential for risk.

## 11.  Make a commitment and empower

In order to achieve real change, the worker needs to define whether they are acting safely or not, and to find a personal reason to start a transformation process that leads them to make safety a lifestyle and not just a procedure of regulatory compliance.

In the final stage of behavioral observation, the observer must pay special attention to the fact that the worker has actually performed an act of conscience, that they are not "caught by surprise" doing something wrong or feeling "reprimanded" for their behavior. For this, it is necessary for the person to clearly see that the way in which they are performing the work is not the safest and the importance of safety.

Here are some suggestions for commitments that you can make with the person approached:

- Make a commitment on the subject dealt with in the approach so that the deviation never happens again;

- Invite the person to be a safety spokesperson;

- Encourage the observed, whenever you find someone committing a behavioral deviation, apply the same tool. At this point you can empower them to practice care as a habit.

## 12. Invest your time genuinely

If any situation requires immediate intervention, do so – even if you have to rearrange schedules. Show your interest in resolving the risk situation, accompany the employee until everything is resolved. Never end an approach by letting the person continue in an imminent risk situation.

## 13. Recognize good practices

When identifying a good practice and/or if something correct is being done, do not hesitate to acknowledge the practice performed. Here are some recognition suggestions:

- Congratulations on the way you are working!

- I watched you working and you inspired me.

- I'm happy to have someone like you in our team.

## 14. Thank them for their attention and time

Point out that the person's time is very important and thank them for paying attention for a few minutes. Never finish an approach without this thank you.

## 15. Complete the behavioral observation form

Use the form to fit your observation into one of these categories of unsafe behavior:

- Use of PPE (absence of use, conservation, and adequacy);

- Procedure (non-compliance, ignorance and lack);

- Human factor (physiological, cognitive, psychological, and social).

Never fill in the form during the approach. It must be completed in a separate place. We know that the form is perceived by the person approached as an audit, something punitive. In addition, we know that people do not audit themselves, we can do this with machines and equipment, but with people, we will need to talk to them. From the conversation we can reflect on their attitudes and build a new behavior and a commitment to life.

## 16. Respect confidentiality

The content of the conversation is confidential, never break this rule.

### 17. **Maintain relationships built**

If you find the one observed by the organization again, do not hesitate to say hello. From now on you have a bond. If you have the opportunity to visit the area again, show that you remember the people you have observed.

## The power of empathy in behavioral observation

Another key element in building a strong safety culture and effective behavioral observation is empathy. Empathy for actively listening and seeing, knowing how to interpret, understand, give voice, inform, demand, hold accountable.

"I must act towards others as I would like them to act towards me" or "I should act towards others as they would like me to act towards them." Which of these two ways should a safety leader think? Undoubtedly, it is much more efficient, in order to resolve risks and ensure safe behavior, for the leader to seek to know how his team thinks, what their desires and dilemmas are. After all, empathy is putting yourself in the other's shoes and allowing yourself to see the world that the other is seeing, feeling what he is feeling, identifying pain, needs and desires, among other aspects.

Empathy allows us to see the other genuinely, putting aside judgments. And when we talk about safety, it is fundamental in the search for answers that reframe limiting beliefs and treat different triggers that trigger important deviations. In addition, the practice of empathy brings intentionality to a leader, makes him sensitive and more respectful of the other. Intentionality, when we talk about leadership, is a mixture of intention, care, sensitivity, and genuine action.

In the search for the regular practice of empathy within organizations, many of them adopt a method called the Empathy Map, a methodology that allows for a deeper understanding of a specific group or individual – be they employees, customers, partners, etc. Created by the American, Dave Gray, the model is an excellent tool for leaders to get to know their teams better and for employees to get to know their colleagues better, creating an environment with an increasing level of empathy.

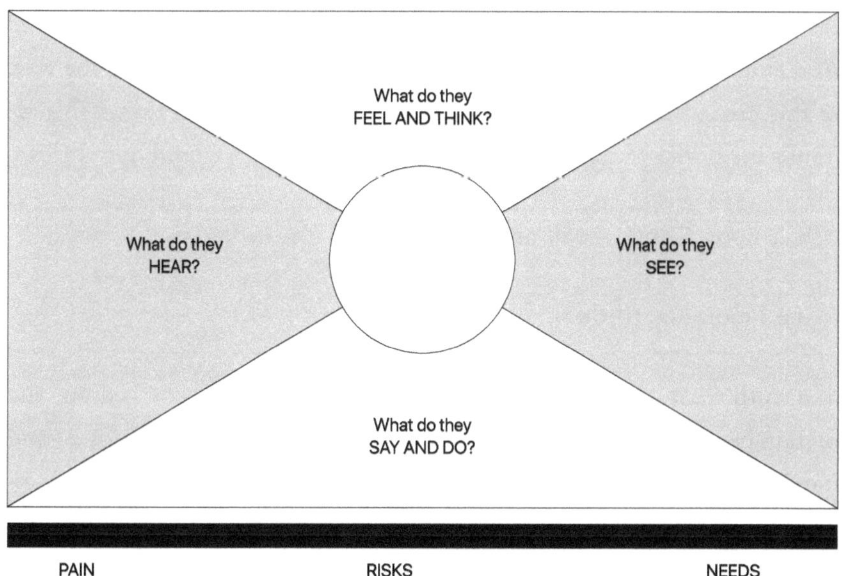


Figure 13: Empathy Map Model

In our very instantaneous days, empathy is an excellent thermometer, able to show you fundamental things, especially when you scroll through the questions on the map without being able to answer them. Every time this happens, which is not uncommon, the leader is forced to look more deeply at himself as a leader, as well as where he has invested his time. He often realizes that not being able to answer the empathy map questions shows that he needs to spend more time with his team. It shows that it is necessary to develop bonds and pave the way for us to be more empathetic.

And empathy is also about accepting the truths that emerge along the way, in this process of a leader listening and getting to know his team. This reminds me of the parable of the truth and the lie, which I reproduce below:

"Once upon a time, the lie and the truth met. The lie said to the truth:

- Good morning, truth.

The truth went to check if it really was a good morning. It looked up, saw no rain clouds and there were birds singing. Seeing that it really was a good morning, it answered to the lie:

- Good morning, lie.

The lie continued:

- It's hot today.

And the truth, realizing the lie was right for the second time, relaxed.

The lie then invited the truth to bathe in the river. It took off its clothes, jumped into the water and said:

- Come on, the water is delicious, really.

As soon as the truth, unsuspecting, took off its clothes and dived in, the lie came out of the water, put on truth's clothes, and left.

Truth, on the other hand, refused to wear lie's garments. And because it had nothing to be ashamed of, it went out naked to walk through the streets and villages. And since then, it's because of this event that, in the eyes of many people, it's easier to accept the lie dressed in truth than the naked truth."

A leader needs to be prepared to see and accept unvarnished truths. Lies dressed up as truth are not the foundation for strong and sustainable safety cultures.

Let's reflect together on one of my favorite dialogues from the book The Little Prince, by Antoine de Saint-Exupéry:

– Good morning – said the fox.

– Good morning – politely replied the little prince, who, looking around, saw nothing.

– I'm here – said the voice, under the apple tree...

– Who are you? – asked the little prince. - You are very pretty...

– I am a fox – said the fox.

– Come play with me – he proposed. - I'm so sad...

– I can't play with you – said the fox. – You haven't captivated me yet.

– Oh, I'm sorry! – Said the little prince.

But upon reflection, he added:

– What does "captivate" mean?

– You're not from here – said the fox. – What are you searching for?

– I am looking for men – said the little prince. – What does "captivate" mean?

– Men – said the fox – have rifles and hunt. That's scary! They raise chickens too. It's the only thing they do that's interesting. Are you looking for chickens?

– No – said the prince. – I look for friends. What does "captivate" mean?

– It is something that is almost always forgotten. It means "to create bonds"...

– Create bonds?

– Exactly. You are nothing to me but a boy just like a hundred thousand other boys. And I have no need of you. And you don't need me either. In your eyes I am nothing but a fox like a hundred thousand other foxes. But if you captivate me, we will need each other. You will be unique to me in the world. I will be the only one for you in the world...

– I begin to understand – said the little prince. – There is a flower... I think it captivated me...

– It's possible – said the fox. - One sees so much on Earth...

– Oh! It wasn't on Earth.

– On another planet?

– Yes.

– Are there hunters on this other planet?

– No.

– That's great! And chickens?

– Neither.

– Nothing is perfect – sighed the fox. – My life is monotonous. I hunt the chickens and the men hunt me. All chickens look alike and so do all men. And that bothers me a little. But if you captivate me, my life will be as if filled with the sun. I will know a sound of footsteps that will be different from the others. The other steps take me underground. Yours will call me out of the hole, as if they were music. And then look! Do you see the wheat fields in the distance? I don't eat bread. Wheat is worth nothing to me. The wheat fields don't remind me of anything. And that's sad! But you have golden hair. Then it will be wonderful when you have captivated me. The wheat, which is golden, will make me remember you. And I will love the noise of the wind in the wheat... – The fox fell became silent and watched the prince for a long time. – Please captivate me! – she said.

– I would like to – said the little prince. – But I don't have much time. I have friends to discover and many things to know.

– We only know well the things we've captivated – said the fox. – Men don't have time to know anything anymore. They buy everything ready-made in stores. But since there are no friend stores, men don't have friends anymore. If you want a friend, captivate me!

– What needs to be done?

– You have to be patient. You will first sit a little far from me, like that, on the grass. And I'll look at you out of the corner of my eye and you won't say anything. Language is a source of misunderstanding. But each day you'll sit a little closer...

The next day the prince returned.

– It would have been better if you had come back at the same time – said the fox. – If you come, for example, at four in the afternoon, from three I'll start to be happy! The more the time comes, the more I will feel happy. At four o'clock, then, I will be restless and agitated: I will discover the price of happiness! But if you come at just any moment, I'll never know when to prepare my heart... There needs to be a ritual.

– What is a "ritual"? – asked the little prince.

– It's a very forgotten thing too – said the fox. – It is what makes a day different from other days; one hour from the other hours. My hunters, for example, love a ritual. They dance on Thursdays with the village girls. Thursday is then a wonderful day! I go for a walk in the vineyard. If hunters danced on any given day, the days would all be the same, and I would never have a vacation!

So the little prince captivated the fox. But when the time for departure came, the fox said:

– Oh! I will cry.

– It's your fault – said the little prince. – I didn't mean to hurt you; but you wanted me to captivate you...

– Yes, I did.

– But you're going to cry! – he said.

– I am – said the fox.

– Then you won't have gained anything!

– I will – said the fox. – Because of the color of the wheat.

Then she added:

– Go and see the roses. Then, you will understand that yours is the only one in the world. You will return to say goodbye to me, and I will present you with a secret.

The little prince went to see the roses:

– You are not the same as my rose, you are nothing yet. No one has captivated you yet, nor have you captivated anyone. You are what my fox was like. She was a fox like a hundred thousand others. But I made her my friend. Now she is unique in the world.

The roses were disappointed.

– You are beautiful, but empty – he continued. – One cannot die for you. Any passerby would no doubt think that my rose looks like you.

She alone is, however, more important than all of you, for she is the one I watered. She was the one I put under the dome. That's where I killed the larvae (except for two or three because of the butterflies). She was the one I heard complain or boast, or even go silent sometimes. Since she is my rose.

And then the fox returned:

– Goodbye! – he said.

– Goodbye. Here's my secret, which is very simple: you can only see well with your heart. The essential is invisible to the eyes. It was the time you wasted with your rose that made it so important. Men have forgotten this truth, but you must not forget it. You become responsible, forever, for what you have captivated. You are responsible for your rose...

– I am responsible for my rose... – repeated the little prince, not to forget.

A classic in literature, the dialogue sums up well the relationship that every leader must have with those they lead, also when it comes to safety culture. What will this culture be like if there is no mutual interest between leader and subordinates, or between the safety team itself and the rest of the corporation? How is it possible to expect an adequate level of safety if there is not an ounce of empathy, admiration, trust, and respect among everyone, without companionship and cooperation? After all, safety culture is about relationships.

The other day I read an interesting piece of information about the importance of developing relationships in the corporate world. Research carried out on the behavior of employees in the work environment increasingly shows the important connection between the management culture of a corporation and the psychological experience that each person who works in that environment reports having. It's a connection that shouldn't go unnoticed within companies, but unfortunately it still does. It seems obvious that the greater an employee's identification with the culture of the company where he works, with what its leaders preach and how they act, the greater will be their psychological well-being. And that people who are connected and engaged in their work feel happier and more satisfied. But, you see, the obvious is not always in the spotlight.

What, exactly, promotes emotional connection and the resulting sense of well-being in team members? A survey carried out in Denmark and published in the Journal of Occupational and Environmental Medicine analyzed 5000 workers and showed that the majority felt emotionally connected with their employers, identified with them, and said they were highly involved with the company where they worked. When analyzing the health of these people, it was found that most slept well and hardly missed work because of illness. The study found that employees with a greater emotional connection to their leaders and work environment not only report a high level of well-being but are also more committed to their colleagues and to the work itself. And, with that, trust between everyone grows.

One day, I was conducting an observation with a supervisor, in his operation, and we found one of his employees inside a machine. The machine, however, was not with Loto. During the approach to this man, the supervisor asked a question that I consider extremely powerful in the process: "Who is waiting for you at home?". The employee promptly replied, "Why do you want to know? We worked together for 5 years and you never asked me."

The supervisor, after a few seconds of embarrassment, replied to the employee: "I really never did ask you, I just realized that today and I would like to start taking better care of you. May I?".

At that moment I wondered why scenes like that happen so rarely. I understand that the most valuable currency we have today is our time, but no result can be sustained without people. They are, in fact, the most precious asset of every leader and every corporation.

## Empathy for the leader

For the leader, it is essential to understand the dimension and importance of empathy. If we know that the leader carries the safety culture, this practice can help him to bring consistency and coherence in the demonstration of a visible commitment to safety.

In addition, his example is powerful and can lead everyone else to experience new habits. The leader must appropriate this tool and create an army of followers willing not only to look around, but also to put themselves in the other person's shoes in search of answers.

In 2017, after news broke that the accident rate at one of his factories was above the industry average, Elon Musk, the powerful executive of Tesla Motors, sent the following email to all of his employees:

"There are no words to express how much I care about your safety and well-being at Tesla. It breaks my heart for anyone to get hurt building cars or trying their best to make Tesla successful.

Therefore, I have requested that all injuries, as of today, be reported directly to me, without exception. I'm meeting with the safety team every week and would like to find all the injured people as soon as they are well so I can understand exactly what we need to do to improve. I will go down to the production line and do the same job they do.

This is what all managers should naturally do. At Tesla, we lead from the front line, not from a safe and comfortable ivory tower. Therefore, our managers must always put the safety of their team above their own."

If Elon Musk really went down to the production line and put himself in the shoes of people who have accidents within his company, this is a great example of how a leader should act. Empathy builds a culture and bond that is hard to break.

## Empathy between colleagues or areas and its application

In daily life and in the practices of safety rituals, empathy must permeate all stages of the process: before behavioral observation, during accident investigations, in the design of projects and safety campaigns...

Empathy is a consequence of relationships and is strongly linked to the corporate culture. And it needs to be practiced as a routine exercise for every leader.

FROM THEORY TO PRACTICE

# Final Words - The New Normal

*"As long as we're united, and as long as we continue to organize, invest, and lead with our values, we'll be unstoppable."*

Tom Perez

I was in London for work, at a mile an hour, participating in yet another certification and thinking about the text to close this book, when I received, in a message, a word that made me reflect: UNSTOPPABLE.

Unstoppable is a person that no one is able to stop, someone with a purpose that, with clarity and simplicity, manages to move forward, always go further. The unstoppable leads a meaningful life, breaks with the status quo and with everything that sounds familiar, is a visitor within the known comfort zones.

What challenges me the most to be unstoppable, every day, is to think about the way someone with this characteristic lives – without reservations, giving the best of themselves, always, without playing for the mediocre team. The unstoppable doesn't wander around full of frustrations or negativity, never talks about what didn't work out or what he isn't able to accomplish. The unstoppable believes.

Sports are full of great unstoppable beings. One of them, who serves as an example to me to this day, never won a single medal. At the 1968 Olympic marathon in Mexico City, Tanzanian John-Stephen Akhwari came in last, more than an hour after the first place. He staggered into the stadium, one leg bleeding from a fall midway through the course. When the audience that still remained in the place spotted him, they were ecstatic and gave him a standing ovation, until he managed to cross the finish line. When asked why he didn't drop out, John replied, "My country sent me to Mexico to finish it." And, without excuses, the runner did. Resilient. Like all those who are unstoppable.

Another story on the topic that I love is based on the book The Star Thrower, by American author Loren Eiseley.

"A young woman was walking along a beach where thousands of starfish had been washed away during a terrible storm. When she reached each star, she would pick it up in her hands and throw it back into the ocean. People watched her with some amusement."

She had been doing this for some time, when a man approached and asked: 'Little girl, why are you doing this? Look around. You can't save all those starfish. There are too many!'

The girl, not looking disheartened, bent down, picked up another star and threw it as far as she could into the water. Then she looked at the man and replied: 'Well, I managed to save another one!'

The old man looked at the girl and thought about what she had done and said. Inspired, he joined the little girl in throwing as many stars as he could back into the sea. Soon others joined them, and more and more stars were saved."

That, for me, is the great value of the unstoppable. In addition to believing, they are able to inspire. In the story about the stars, it would be normal to think that it would not be possible to save that huge amount lying on the sand. But the unstoppable, with their purpose and resilience, are capable of creating the NEW NORMAL.

We live in an era full of new normals. New records, new brands, new services, new ways of thinking, new technologies, new businesses. Normals who are driven by purpose – and be aware that purposes are not mere words in the wind, but something that needs to be embraced and lived.

In the era of the NEW NORMAL, we live in a new status quo, a new ruler, a new mindset. And organizations, as well as the people who are part of them, need to be prepared for this scenario.

I still find, during the work I do in the safety area, many people resistant to this NEW NORMAL. Some just don't accept change, don't see value in a continuous improvement process, don't accept good ideas from other people. Others accommodate because of bureaucracy, find the minimum necessary sufficient, deliver only what is asked of them, are unable to collaborate.

There are also those who cling to past experiences without looking forward. There are those who are negative, now known as haters, for whom news is never welcome. And there are those who are insecure, who do not believe in their own knowledge and potential to experience change. Because of fear, they become paralyzed and are only willing to see the NEW NORMAL if they are 100% sure of it.

To all these people, I hope this book has served as fuel or at least a spark for change. May we build a solid and sustainable process of creating and strengthening culture. I hope this book has presented enough information for that. Now the main element is missing: YOUR PERSONAL AND GENUINE WILL TO WANT TO BE BETTER, ALWAYS.

## Big hug,

Andreza Araújo

# Limiting Beliefs

- It's always been like this when I arrived they already did things exactly like this.

- My work has no risks

- The hazards and risks are in the factory, not here in the office

- Here in the department we do it like this.

- Everyone does it that way.

- Nothing ever happened here.

- The speed limits for this highway were established 20 years ago.

- Doing it safely costs more.

- We are in a hurry, I needed to buy time.

- I just got back from vacation, I'm warming up, calm down…

- I pass by this street every day.

- Nobody knows about this procedure; it only exists at management.

- Safety personnel are not here right now.

- If we don't, we will lose our customers.

- I thought someone checked first.

- ✔ Someone must be taking care of it.

- ✔ I found a way to make this activity faster.

- ✔ It has always been like this when we do this activity we can skip these steps.

- ✔ All ready to buy time.

- ✔ Ah, we only do this when we have an audit.

- ✔ I know this path with my eyes closed.

- ✔ Do you know how long I've been doing this?

- ✔ Do you know how long I've been working here?

- ✔ I've seen just about everything here.

- ✔ If you forget about the procedure and do it safely, it will take longer.

- ✔ We made a technical tweak here and now it's faster.

- ✔ I think if I was insecure someone would have already said something.

- ✔ Here in the department we decided to do this.

- ✔ The legislation is greatly exaggerated.

- ✔ This factory is very old, so we were unable to define and enforce safety procedures.

- ✔ These equipment are very robust; I think we can extend and delay preventive maintenance a bit.

- This guy is a former employee, knows everything around here and is very committed to us (trustworthy).

- You can go in reverse; nobody ever stops here.

- Do not stop at traffic lights, it's very dangerous. Accelerate!

- I know all the faults of this equipment.

- I don't remember the contents of the last DDS.

- We took advantage of the safety committee meeting to talk about the Monday team's vacation.

- For 500 days we've had no accidents with leave of absence.

- Everyone knows the golden rules, I find it very repetitive to use this content, it's already very outdated.

- We are not finding opportunity to improve on our behavioral observations. We are only doing positive reinforcement.

- I visited the factory today and I don't see any improvement opportunities for safety.

- I'm feeling a false sense of safety.

- I see no need to repeat this training this year.

- We already know our tasks.

- Almost no accidents happen here.

- We met all of our safety goals.

- I believe we have already reached a level of excellence in safety.

- This job is so simple that anyone can do it.

- The hazards and risks are in the factory, here in the office nothing happens.

- My work is risk free.

- What if it happened? Stop that kind of talk! This is super controlled. Talk to the safety guy and confirm it.

- This work is so important that, for it to be safer, we are asking the safety area to follow up.

- I can do and sign routine service releases right from my office.

- A trained employee is a qualified employee.

- If nobody could walk through here, this area would be closed and/or partially closed.

# Leadership Assessment

## What Senior Leadership Says About Safety

| Topic | Pathological | Reactive | Calculating/Bureaucratic | Proactive | Sustainable |
|---|---|---|---|---|---|
| **Talk & Do** | 1– I accept that it's possible for some accidents to happen and that doing it safely is sometimes impossible. | 2– Meeting safety targets is important to achieving the bonus. If I have to deliver a result, I negotiate safety. | 1– It's more important to say that safety is important than having Safety.<br><br>2 – I comply with some safety procedure. | 1– I make visits focused on Safety, I'm proud of the results we have achieved in recent years. My personal goal is for us to achieve excellence in Safety. | 1– I feel uncomfortable when I feel surprised by new Safety demands.<br><br>2– My desire is to influence society with practical demonstrations.<br><br>3– I want to constantly challenge people not to rely on luck. |
| **Leadership Attributes** | 1– After an accident, the important thing is to find a culprit and ensure that it cannot be avoided. Employees need to feel comfortable carrying out rates the way they have been taught. | 1– I am disciplined in applying the Progressive Discipline policy, but I don't believe it should be used for leadership, nor do I believe that we need to fire someone who breaks a Safety rule. | 1– I see so many Safety campaigns, I don't understand why accidents keep happening<br><br>2– People tell me they are doing it safely. | 1– I question the action plans and closure of investigations. I trust that my team is evolving in the search for excellence in Safety. | 1– Everyone can talk to me openly about Safety.<br><br>2– My role supports my team so that they do it safely and I always and constantly share my vision of Safety with the entire organization and their families. I seek to build a Safety coalition. |

| Topic | Pathological | Reactive | Calculating/ Bureaucratic | Proactive | Sustainable |
|---|---|---|---|---|---|
| **Decision Making** | 1- My priority is to ensure the organization's financial results. Safety is everyone's responsibil ty. | 1- I strive my organization to avoid loss and damage in Safety. I look forward to accident classification. My visits reflect the serious problems we have to solve in Safety. | 1- I try to show my displeasure with bad results in Safety, I always have a current message on the subject. That's all I can do. | 1- I seek to achieve excellence in Safety among my peers.<br><br>2- I value the sharing of Lessons learned, it irrigates my knowledge. Personally, Safety means care. | 1- I resist making concessions in my Safety decision making.<br><br>2- I do not allow projects to be conceived without the participation of Safety. |
| **Knowledge** | 1- I believe my role is to tell people to use the PPE we deliver and work safely.<br><br>2- The discourse and plans of the Safety area do rot attract.<br><br>3- We still need to hide some issues, because Brazilian legislation s very exaggerated. | 1- I believe that the Safety area should take care of Safety issues.<br><br>2- They must present and conduct the topic here in our company. A very technical topic. | 1- It's important that we get all Safety certifications.<br><br>2- It's important to keep zero accidents with leave of absence and expand the records of days, I consider myself well. | 1- I always get involved in incident investigations to ensure we get to the root cause. Safety success is the base of pyramid indicators. I engage my operation in the search for deviation and near misses. I admit I have a lot to learn. | 1- I enjoy learning about Safety. I reinforce the importance of a transparent database, where all occurrences are reported and investigated. I challenge people not to feel a false sense of safety and ensure that everyone can identify deviations and unsafe conditions. |

# Bibliography


**BOOKS**

ARAUJO, Andreza. Make a Difference: Be a Leader in Health and Safety. Nelpa, 2014.

ARAUJO, Andreza. Safety Leadership Practical Guide. Nelpa, 2014.

ARAUJO, Giovanni M. de. SMSQRS Management System Elements - Vulnerability Theory - Volume 1. Verde Editora Management, 2009.

COOPER, Dominic. Improving Safety Culture: a Practical Guide. Wiley, 1997.

GOLDSMITH, Marshall. The Trigger Effect: How to Trigger the Behavior Changes That Lead to Success in Business and Life. Companhia Editora Nacional, 2017.

HOLANDA, Sérgio Buarque de. Brazilian roots. Companhia das Letras, 2015.

KAHNEMAN, D., SLOVIC, P., TVERSKY, A. Judgment Under Uncertainty: Heuristics and Biases. Cambridge University Press, 1982.

REASON, James. Managing the Risks of Organizational Accidents. Routledge, 1997.

ROACH, Marie S. The human act of caring: a blueprint of the health professions. Canadian Hospital Association, 1993.

SAINT-EXUPÉRY, Antoine de. The Little Prince. Gallimard, 1943.

SHAKAROV, Andrei. Memoirs. Random House Value Publishing, 1995.

## ARTICLES

GULDENMUND, F. W. The Nature of Safety Culture: A Review of Theory and Research. Safety Science Group, Elsevier, 2000. At www.sciencedirect.com/science/article/pii/S092575350000014X?via%3Dihub

JICK, T.D. Mixing Qualitative And Quantitative Methods: Triangulation in Action. Administrative Science Quarterly, 1979. At www.jstor.org/stable/2392366?origin=crossref&seq=1#page_scan_tab_contents

KAPITZA, Sergei P. Lessons of Chernobyl: The Cultural Causes of the Meltdown. At www.foreignaffairs.com/articles/russian-federation/1993-06-01/lessons-chernobyl-cultural-causes-meltdown

First edition Portuguese (November 2019)

Second edition in English (June 2022)